PEARSON MATH

ALGEBRA 1
GEOMETRY
ALGEBRA 2

Teaching With TI Technology

Laurie E. Bass, Allan E. Bellman, Art Johnson

Boston, Massachusetts
Chandler, Arizona
Glenview, Illinois
Upper Saddle River, New Jersey

ISBN-13: 978-0-13-370608-6
ISBN-10: 0-13-370608-7
3 4 5 6 7 8 9 10 V084 13 12 11 10

Teaching With TI Technology

Contents

Introduction

Activities and the compact disc (CD) in this booklet accompany lessons from the Prentice Hall textbooks *Algebra 1, Geometry,* and *Algebra 2,* and *CME Project* Algebra 1, Geometry, and Algebra 2.

The Activities have been designed for use with Texas Instruments (TI) handheld technology, specifically the TI-84 Plus, TI-84 Plus Silver Edition, TI-83 Plus, and TI-83 Plus Silver Edition handheld devices. (In this booklet, "TI-83/84 Plus" refers to all four devices.) The Activities also use TI Apps and Prentice Hall programs.

Apps TI's Handheld Software Applications (Apps) provide added functionality, such as the ability to construct geometric figures.

Programs The Prentice Hall programs provide further enhancements to functionality. A special type of program called an *AppVar* can run only when its corresponding App is running.

Activities and Teacher Notes

- You can use the Activities in this booklet (and on the accompanying CD) to introduce, investigate, and extend the ideas of the lessons. Students can work individually with their calculators, but you can invite them to share discoveries as they proceed.
- You can use the Teacher Notes for each Activity to find out more about the Activity, including answers to Activity exercises.

The Accompanying CD

The CD, in the sleeve inside the back cover, includes:

- Activities and Teacher Notes from this booklet in print-ready form.
- Apps and programs (including AppVars) needed for the Activities.
- A User's Guide for each App.
- A *TI Technology Guide* for detailed information about the TI Apps and related technology.
- TI Connect™ software, which allows you to work with items on the CD and lets your computer "talk" with your TI-83/84 Plus.
- Links to ready-to-play instructional videos powered by TI-SmartView™.
- A 365-day trial of TI-SmartView™.

Starting an Activity

What You Need

- this booklet
- the CD on the inside back cover
- TI-83/84 Plus handheld devices for you and your students
- TI cable for computer to calculator transfers and connectivity cable(s) for calculator to calculator transfers

Before You Begin

Install the CD and TI Connect on your computer.

Select an Activity from the workbook, and note the files listed at the top of the Activity page.

Launching an Activity

1. Connect your computer and your TI-83/84 Plus using the TI cable.
2. Run TI Connect. Open the TI Device Explorer.
3. Run the Teaching with TI Technology program. Click on the Activities tab, then choose your subject: *Algebra 1*, *Geometry*, or *Algebra 2*. Locate your file name and drag or copy it into the TI Device Explorer window.
4. Distribute the files to each TI-83/84 Plus in the classroom with connectivity cables or TI-Navigator™.
5. Do the Activity.

Prentice Hall Algebra 1 and Topics in TI Algebra 1 Apps

Activities for Chapters 1–6 of the textbook *Prentice Hall Algebra 1* are built into the Topics in Algebra 1 Apps. Use the guide below to find the activities for the TI-83/84 Plus. This guide will also help you locate the related worksheets and Teacher Notes in the *Topics in Algebra 1* manuals on the CD.

***Algebra 1* Lesson**		**App**	**Section(s)**
1-3	Real Numbers and the Number Line	ALG1CH1 (Number Sense)	Integers, Rational Numbers, Real Numbers
1-4	Properties of Real Numbers	ALG1CH1 (Number Sense)	Real Numbers
1-5	Adding and Subtracting Real Numbers	ALG1CH1 (Number Sense)	Integers, Rational Numbers, Real Numbers
1-6	Multiplying and Dividing Real Numbers	ALG1CH1 (Number Sense)	Integers, Rational Numbers, Real Numbers
1-7	The Distributive Property	ALG1CH1 (Number Sense)	Real Numbers
2-1	Solving One-Step Equations	ALG1CH2 (Linear Equations)	Using Graphs & Tables Using Algebra
2-2	Solving Two-Step Equations	ALG1CH2 (Linear Equations)	Using Algebra
3-1	Inequalities and Their Graphs	ALG1CH4 (Linear Inequalities)	Using Graphs & Tables
3-2	Solving Inequalities: Using Addition or Subtraction	ALG1CH4 (Linear Inequalities)	Using Graphs & Tables Using Algebra
3-3	Solving Inequalities: Using Multiplication or Division	ALG1CH4 (Linear Inequalities)	Using Algebra
4-4	Graphing a Function Rule	ALG1CH2 (Linear Equations)	Using Graphs & Tables
4-6	Formalizing Relations and Functions	ALG1CH2 (Linear Equations)	Using Graphs & Tables
5-1	Rate of Change and Slope	ALG1CH3 (Linear Functions)	Slope With Grid, Slope Using Coordinates, Slope as Rate of Change, Slope-Intercept Form
5-3	Slope-Intercept Form	ALG1CH3 (Linear Functions)	Slope-Intercept Form
6-1	Solving Systems by Graphing	ALG1CH5 (Linear Systems)	Using Graphs & Tables
6-2	Solving Systems Using Substitution	ALG1CH5 (Linear Systems)	Using Algebra
6-3	Solving Systems Using Elimination	ALG1CH5 (Linear Systems)	Using Algebra
6-4	Applications of Linear Systems	ALG1CH5 (Linear Systems)	Using Graphs & Tables Using Algebra

CME Project Algebra 1 and Topics in TI Algebra 1 Apps

Activities for Chapters 1–6 of the textbook *Prentice Hall Algebra 1* are built into the Topics in Algebra 1 Apps. Use the guide below to find the activities for the TI-83/84 Plus. This guide will also help you locate the related worksheets and Teacher Notes in the *Topics in Algebra 1* manuals on the CD.

***Algebra 1* Lesson**		**App**	**Section(s)**
1.2	Thinking about Negative Numbers	ALG1CH1 (Number Sense)	Integers, Real Numbers
1.5	The Basic Rules of Arithmetic	ALG1CH1 (Number Sense)	Real Numbers
1.13	Adding and Subtracting Fractions	ALG1CH1 (Number Sense)	Rational Numbers, Real Numbers
1.15	Multiplying and Dividing Fractions	ALG1CH1 (Number Sense)	Rational Numbers, Real Numbers
2.11	The Basic Moves for Solving Equations	ALG1CH2 (Linear Equations)	Using Algebra
2.12	Solutions of Linear Equations	ALG1CH2 (Linear Equations)	Using Algebra
4.2	Pitch and Slope	ALG1CH3 (Linear Functions)	Slope With Grid, Slope Using Coordinates
4.3	Rates of Change	ALG1CH3 (Linear Functions)	Slope with Grid, Slope Using Coordinates, Slope as Rate of Change
4.7	Jiffy Graphs: Lines	ALG1CH3 (Linear Functions)	Slope-Intercept Form
4.9	Getting Started	ALG1CH5 (Linear Systems)	Using Graphs & Tables
4.10	Solving Equations: Substitution	ALG1CH5 (Linear Systems)	Using Algebra
4.12	Solving Equations: Elimination	ALG1CH5 (Linear Systems)	Using Algebra
4.14	Inequalities in One Variable	ALG1CH4 (Linear Inequalities)	Using Algebra
6.10	Rational and Irrational Numbers	ALG1CH1 (Real Numbers)	Rational Numbers, Real Numbers

TI Technology Guide at a Glance

The CD includes a complete *TI Technology Guide*. Here are some *Guide* topics in brief. The *Guide* gives details.

TI Connect™ Software

You install the TI Connect software on your Windows® PC or Macintosh® computer. In addition to letting you transfer files between your computer and your TI-83/84 Plus, this software enables you to view or save TI-83/84 Plus files and screen images on your computer, paste screen images into other applications, and make, display, and modify list files, number files, and matrix files.

TI-83/84 Plus Operating System

At the start of the school year, check that each TI-83/84 Plus has the current operating system (OS). With your TI-83/84 Plus connected to your computer and your computer connected to the Internet, run TI Connect. On the TI Connect screen, click on the update button. If you find that you need to update, follow the on-screen instructions.

Finding the TI-83/84 Plus Files You Need

The names of the Apps (see next page) suggest their use. The names of the programs (including AppVars) and data lists are coded, but the codes are keyed to the intended use. For example, the program A1L66B (see page 11) accompanies *Prentice Hall Algebra 1* Activity for Lesson 6-5. It's the second such program, after A1L66A.

Running an App, Program, or AppVar

For an App, press **APPS**. Select the App and press **ENTER**.

For a program (that is not an AppVar) used with an App, have the App running. Press **PRGM**. Select the program and press **ENTER** twice.

For a Cabri® Jr. AppVar, have Cabri Jr. running. Select Open in the Cabri Jr. F1 menu. Press **ENTER**, select the AppVar, and press **ENTER**.

For a StudyCards™ AppVar (stack), have the StudyCards App running. In the MAIN MENU, select 2: CHOOSE NEW STACK. Select the desired stack and press **ENTER**.

Note: Each Activity assumes that you have the appropriate App running from the start. If one App is active and you try to run another, you may receive an error message that prompts you to choose between the Apps.

Teaching With TI Technology Apps on the CD

Here are the TI Apps available on the CD and a brief description of each.

Cabri® Jr. (CabriJr)	provides interactive, dynamic geometry modeling. Learn more about this App on pages x–xi.
Guess My Coefficients App (GuesCoef)	challenges students to identify the coefficients of linear, quadratic, and absolute value functions from characteristics of their graphs.
Inequality Graphing App (Inequal)	simplifies the graphing of inequalities and lets you store the coordinates of special points.
Polynomial Rootfinder and Simultaneous Equation Solver App (PolySmlt)	finds the roots (zeros) of polynomials of degree 1 through 30. It also finds solutions of systems of linear equations, or tells you that a given system has no solution or infinitely many solutions.
Probability Simulation App (Prob Sim)	simulates various games of chance for probability investigations, generates random numbers, and provides ways to summarize data.
Science Tools App (SciTools)	calculates with significant digits, converts measurements and scientific constants from one unit to another, provides ways to display data and find best-fit functions for given data, and performs basic vector operations.
Solve it! App (SolveIt)	provides practice on algebra problems at different skill levels.
StudyCards™ App (StudyCrd)	displays electronic flash cards to review key terms and concepts. Learn more about this App on page xiv.
Topics in Algebra 1 Apps	provide tutorial and practice activities for early algebra 1 chapters. Learn more about this App on pages vi–vii.
Transformation Graphing App (Transfrm)	shows and uses the effects of changing parameter values on graphs of functions. Learn more about this App on page xii.

Your CD contains a complete User's Guide for each App.

StudyCards Creator

The CD also contains the TI desktop tool, StudyCards Creator for Windows PC only. It lets you make your own StudyCards stacks or change existing stacks.

Memory

Like any computer, your TI-83/84 Plus has pre-set memory capacity, with the amounts varying among the TI-84 Plus, TI-84 Plus Silver Edition, TI-83 Plus, and TI-83 Plus Silver Edition. If you try to load more than the device can hold, you will get a memory error. To learn more about memory capacity, see the *TI Technology Guide* on the CD.

Cabri® Jr.

The Cabri® Jr. application lets you construct, analyze, and transform mathematical models and geometric diagrams. Cabri Jr. was developed for TI by Cabrilog and renowned French mathematician Jean-Marie Laborde.

With Cabri Jr. you can:

- Perform analytic, transformational, and Euclidean geometric actions.
- Make geometric constructions interactively with points, lines, polygons, circles, sets of points for loci, and other basic geometric objects.
- Alter geometric objects in real time to see patterns, make conjectures, and draw conclusions.

Hints and Tips

In Cabri Jr., you can access the F1–F5 menus by pressing the keys Y=, WINDOW, ZOOM, TRACE, GRAPH just below the TI-83/84 Plus screen. You may wish to keep this page in front of you until you become familiar with them.

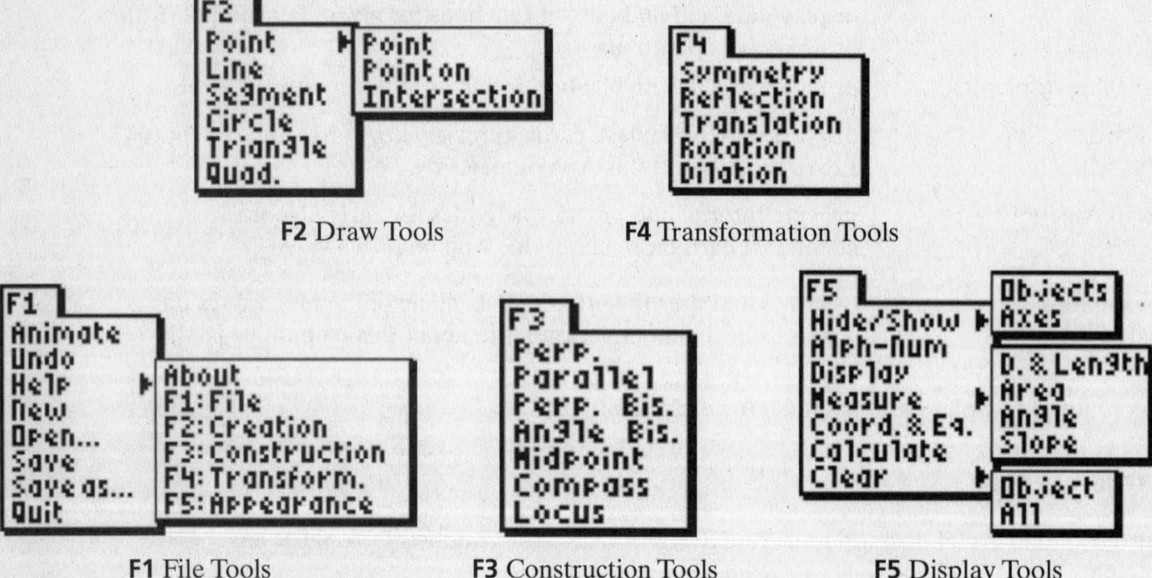

F2 Draw Tools

F4 Transformation Tools

F1 File Tools

F3 Construction Tools

F5 Display Tools

The F1 Menu

Animate — Select Animate and press ENTER. Select a point that was constructed on a segment or circle. When you see the double arrow, press ENTER.

You can stop an animation in three ways:

1. When the animated point returns to the cursor and you see the double arrow, press ENTER.
2. Press 2nd and then ENTER to stop all animated points.
3. Select Undo and press ENTER.

Undo — Select Undo and press ENTER.

Open... — Select Open and press ENTER to see the AppVar list.

The F2 Menu

Point Select Point and press ENTER. Move the cursor to a desired location and press ENTER. To construct a point on an object, be sure the object is flashing before you press ENTER. To construct a point at the intersection of two objects, be sure both objects are flashing.

Line or Segment Select Line or Segment and press ENTER. Move the cursor to a desired location for one point and press ENTER. Repeat for a second point.

Circle Select Circle and press ENTER. Move the cursor to a desired location for the center and press ENTER. Move the cursor to a desired point on the circle and press ENTER.

Triangle or Quadrilateral Select Triangle or Quadrilateral and press ENTER. For each vertex, move the cursor to the desired location and press ENTER.

The F3 Menu

Parallel Select Parallel and press ENTER. Select a line or segment and press ENTER. Use the arrows to drag the parallel line to the desired location.

To construct a parallel line through a given point, select Parallel and press ENTER. Select the point and press ENTER. Select the line (or segment) and press ENTER.

The F5 Menu

Measure Select Measure and then select D & Length, Area, Angle, or Slope from the submenu. Press ENTER. Select the object to be measured. Press ENTER to install the measurement.

Use the arrows to relocate an installed measurement and press ENTER.

Use D & Length to measure a side or the perimeter of a triangle or quadrilateral. Move the cursor to the figure so that the side (for length) or the entire figure (for perimeter) is flashing. (Press 2nd to switch which is flashing.) Then press ENTER.

Clear Select Clear and then select Object or All from the submenu. Clear Object gives you an erasing cursor. To erase an object, move the cursor to the object and press ENTER. Clear All erases everything on the screen.

You can erase some objects without using Clear. With no tool active, select the object. If the cursor becomes hollow, press DEL.

You can also erase the screen. Press CLEAR three times and then press ENTER.

Screen Mechanics

To change or deactivate a tool, press CLEAR.

To drag an object, deactivate the current tool and move the cursor to the desired object. If the cursor becomes hollow, press ALPHA. Drag the object with the hand-shaped cursor that appears. To stop dragging, press ENTER or CLEAR.

Transformation Graphing App

Use the Transformation Graphing App to investigate how changing the parameter values A, B, C, and D affects the graphs of these functions.

Y = AX + B	Y = A |X − B| + C	Y = (X − A)(X − B)(X − C)
linear	**absolute value**	**polynomial**

Y = AX² + C   Y = AX² + BX   Y = AX² + BX + C   Y = A(X − B)² + C

quadratic (various forms)

Y = A√(X − B) + C	Y = AB^X + C	Y = (A/(X − B)) + C	Y = A sin(B(X − C)) + D
square root	**exponential**	**rational**	**trigonometric**

Hints and Tips

Name Press APPS. The Transformation Graphing App is named Transfrm.

Running You can tell whether the Transformation Graphing App is active by checking the Y= screen. The App is active if you see special icons in front of Y1, Y2, etc. Note that you can select only one function at a time.

Quitting To quit the Transformation Graphing App, press APPS, select Transfrm, and press ENTER. You will see a menu with the choices 1: Uninstall and 2: Continue. Select Uninstall and press ENTER. (*Note*: This does not remove the App from your TI-83/84 Plus.)

Parameter Values Programs used with the Transformation Graphing App often show parameter values on the graph screen. To change a parameter value, select the parameter by highlighting its = sign using the up and down arrow keys. Type the new value and press ENTER.

Speed Tip: Press the right (left) arrow key to increment (decrement) parameter values by the Step value. (See **Settings**.)

Settings (Play and Play-Fast) Press WINDOW and the up arrow key to see the SETTINGS menu. This menu gives you a way to change the Step value and another way to change the screen parameter values. (See **Parameter Values**.)

The SETTINGS menu also shows three icons, >||, >, >> that represent Play-Pause, Play, and Play-Fast, respectively. If you select Play or Play-Fast and press ENTER, Max also appears. Use Max to set the maximum value for the selected parameter.

Animate Once you have set the values for the parameters and Max (see **Settings**), press GRAPH to see a "slide show." The screen will cycle through graphs for values of the selected parameter from its initial value through the incremental steps to the Max value.

The slide show can show at most 13 graphs. If the screen shows a memory error, change the initial value of the parameter, the value of Max, or the size of the step to reduce the number of graphs.

To stop the slide show, press ENTER. To resume, press ENTER again. Press ON to go directly to the SETTINGS menu.

Inequality Graphing App

Use the Inequality Graphing App to streamline the graphing of inequalities, including graphs with vertical boundary lines. For linear programming activities, you can show the region of feasible points, trace on and store the coordinates of corner points and other points of interest, and evaluate objective functions at all such points.

Hints and Tips

Name Press APPS. The Inequality Graphing App is named Inequal.

Graphing On the Y= screen, enter an equation or view an equation entered by running a program. Select the = sign. Press ALPHA F2, F3, F4, or F5 for $<, \leq, >,$ or $\geq$, respectively. Then press GRAPH.

Shading This App shades the solution region for each inequality. To shade an intersection (the solution of a system) or a union, use Shades shown on the graph screen. Press ALPHA F1 (or F2) to see the choices.

Quitting To quit the Inequality Graphing App, press APPS, select Inequal, press ENTER, select Quit Inequal, and press ENTER.

Using DEFAULT

DEFAULT is a general program that clears the entries from the Y= screen. It also restores the standard viewing window and various general mode settings. You can load DEFAULT into the class handhelds at the beginning of your Algebra 1 or Algebra 2 course and use it throughout the year. It is good practice to run DEFAULT after you finish using any program within the Transformation Graphing or Inequality Graphing Apps.

DEFAULT does not restore = signs that you replaced in the Y= screen. See **Graphing** above.

StudyCards™ App

StudyCards™ software lets students replace the traditional practice/review stacks of index cards with electronic "cards" they can view on their TI-83/84 Plus.

You can-make stacks of cards using TI's StudyCards™ Creator software (Windows PC only). Students can view these stacks on their TI-83/84 Plus handhelds using the StudyCards™ App.

Two typical cards are shown below:

Front side:

```
The distance of a number
from 0 on a number line is
the __ ? __ of the number.

MENU|YES| NO |FLIP|STAT
```

FLIP side:

```
absolute value

MENU|YES| NO |FLIP|STAT
```

Front side:

```
What must be true about
numbers a, b, and c for
you to conclude ac < bc?

1) a > b, c > 0
2) a < b, c < 0
3) a > b, c < 0
4) a < b, c = 0
MENU|CHOICE:?|FLIP|STAT
```

FLIP side:

```
The correct answer is
3) a > b, c < 0.

By the Multiplication
Property of Inequality,
if a > b and c < 0, then
ac < bc.
MENU|CHOICE:3|FLIP|STAT
```

The basic steps for making a stack of cards are shown below.

1. Open StudyCards™ Creator on your computer.
2. Select **Self-Check** in the New Stack dialog box.
3. Click **OK**.
4. Enter the **Stack tile**, **Version no.**, and **Created By** information on the top half of the screen.
5. In the **Card Info** section, complete the following steps for each new card:
   a. Enter the **Name of Card**.
   b. Place cursor in the **Front of Card** field and type text.
   c. Place cursor in the **Back of Card** field and type text.
   d. Select **New Card** to make the next card, if desired.
6. Select **Stack > Save** to save the stack.
7. Select the file type based on the type of handheld students will use to view the cards
8. Click **save**.
9. Transfer the stack to your calculator(s).

For more details see the StudyCards™ Creator and StudyCards™ App User's Guides on the CD that accompanies this booklet.

Name ____________________ Class ____________ Date ____________

Measurement Conversions

Activity 1

FILES NEEDED: Science Tools App

In the Science Tools App, you can use the UNIT CONVERTER to switch from one unit to another. Follow this example to convert 2 meters to feet.

In the UNIT CONVERTER, find and open the LENGTH menu.

SELECT A TOOL
1:SIG-FIG CALCULATOR
2:UNIT CONVERTER
3:DATA/GRAPHS WIZARD
4:VECTOR CALCULATOR
EXIT

UNIT CONVERTER
1:LENGTH 7:MASS
2:AREA 8:FORCE/WT
3:VOLUME 9:PRESSURE
4:TIME A:ENERGY/WORK
5:TEMP B:POWER
6:VELOCITY C:SI PREFIXES
CONSTANT

Type 2, the number of meters you want to convert.

LENGTH
fm Å mm cm m
km Mil in ft yd
fath rd mi nmi ltyr
2
CONSTANT EXPT COPY EDIT

Choose the given unit of measure and press ENTER.

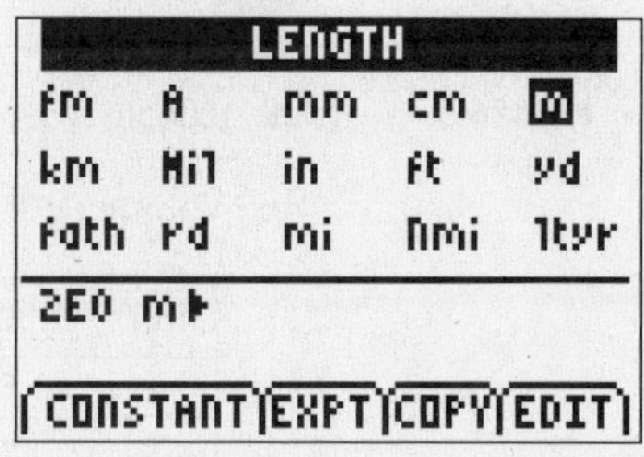

Choose the desired unit of measure and press ENTER.

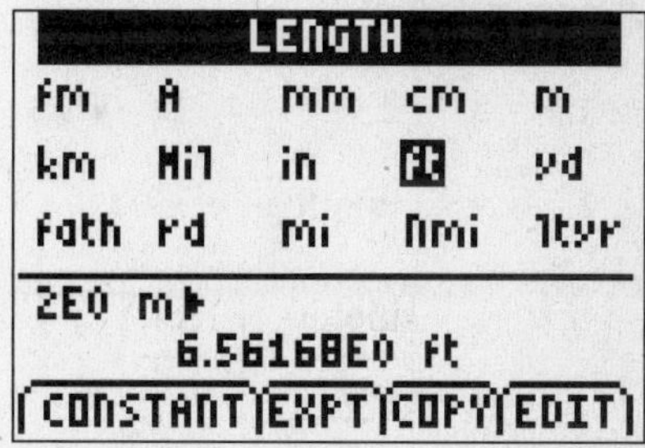

The screen shows 2 m ≈ 6.56 ft.

1. The converted value appears in a calculator form of scientific notation. Write the standard form for each number given in calculator scientific notation below. (If scientific notation is new to you, look ahead to Lesson 8-2.)

a. 1.23E1 **b.** 1.23E–1 **c.** 4.56E2 **d.** 4.56E–2

2. Use the UNIT CONVERTER to complete the table. State whether the converted values vary directly with the given measurements. If they do, write an equation for the direct variation.

a. Length

meters	feet
2	6.56
3	
4	
6	

b. Volume

liters	gallons
2	
3	
4	
6	1.59

c. Temperature

°C	°F
2	
3	
4	
6	

d. Volume

cups	gallons
2	
3	
4	
6	

Measurement Conversions

Activity Objective

Students use the Science Tools App to convert measurements.

Time

- 15–20 minutes

Materials/Software

- App: Science Tools
- Activity worksheet

Classroom Management

- Students can work individually or in pairs depending on the number of calculators available.

Notes

- Students may need to review scientific notation before answering Question 1.

Answers

1. **a.** 12.3 **b.** 0.123 **c.** 456 **d.** 0.0456

2. **a.** Length; $y = 3.28x$

meters	feet
2	6.56
3	9.84
4	13.12
6	19.69

b. Volume; $y = 0.265x$

liters	gallons
2	0.53
3	0.79
4	1.06
6	1.59

c. Temperature; not direct variation

°C	°F
2	35.6
3	37.4
4	39.2
6	42.8

d. Volume; $y = 0.0625x$

cups	gallons
2	0.13
3	0.19
4	0.25
6	0.38

Name ____________________ Class ____________ Date ____________

Linear Graphs

Activity 2

FILES NEEDED: Transformation Graphing App
Program: A1L62A

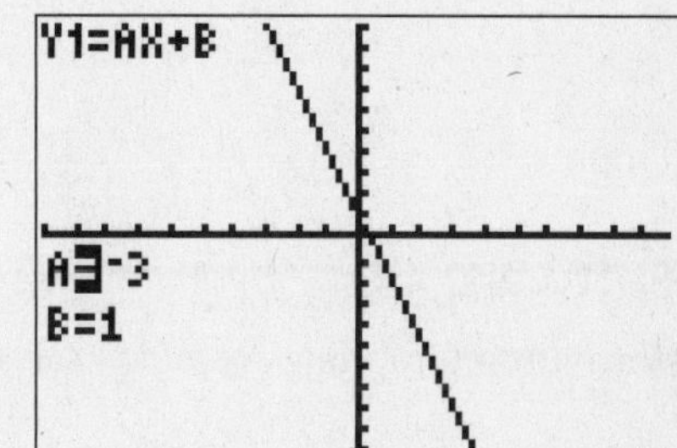

A1L62A shows the linear equation or function $y = mx + b$ graphed as Y1 = AX + B.

In this activity, you will explore the graph of $y = mx + b$.

1. Run A1L62A. Increment the values of A in steps of 1. How does the graph change?
2. Predict how the graph will change when you increment the values of A in steps of −1. Test your prediction.
3. Increment the values of B in steps of 1. How does the graph change?
4. Predict how the graph will change when you increment the values of B in steps of −1. Test your prediction.
5. Set A = 0 and B = 5. What kind of line do you see?
6. Predict what will happen for A = 5 and B = 0. Test your prediction.

Extension

7. Is it possible to make the line vertical by adjusting the values of A and B? Explain.
8. If A1L62A had set up the graph of Y1 = A + BX instead of Y1 = AX + B, how would the results for the activities above differ? Explain.

Animation Option

Set the Transformation Graphing App and select Play [>] as shown at the right. Set A = −3, B = 1, Step = 1, and Max = 5. Select A for incrementing.

9. Press GRAPH and watch the values of A increase from −3 to 5. Does the line move the same amount with each change? Explain.
10. Select B for incrementing. Before you press GRAPH, predict what you will see.
    Press GRAPH and check your prediction.

Linear Graphs

Activity Objective

Students use the Transformation Graphing App to explore the graph of the function $y = mx + b$.

Time

- 15–20 minutes

Materials/Software

- Transformation Graphing App
- Program: A1L62A
- Activity worksheet

Skills Needed

- start an App

Classroom Management

- Students can work individually or in pairs depending on the number of calculators available.
- Use TI Connect™ software, TI-GRAPH LINK™ software, the TI-Navigator™ system, or unit-to-unit links to transfer A1L62A to each calculator.

Notes

- Remind students that Y1 = AX + B is screen notation for $y = mx + b$, where A = m.
- Some values in the Play SETTINGS may lead to errors. The number of screens is determined by how many times the selected value can be incremented without exceeding the Max value.

Answers

1. The line tilts upward to the right.

2. The line will tilt downward to the right.

3. The line moves up but its tilt doesn't change.

4. The line will move down.

5. a horizontal line through (0, 5)

6. a steep line through (0, 0)

7. No; as the value of A get larger and larger, the line gets closer to vertical but will not reach it.

8. The answers for Questions 1–6 would be the answers above for Questions 3, 4, 1, 2, 6, 5, respectively. The answer for 7 would be the answer above for 7 with A replaced by B.

9. No; the line moves the most when A changes from -1 to 0 and from 0 to -1.

10. The line will move up.

Name ________________________ Class ______________ Date ______________

Linear Function Match

Activity 3

FILES NEEDED: Transformation Graphing App, Guess My Coefficients App
Program: A1L62B

A1L62B graphs the linear equation or function $y = mx + b$ as Y1 = AX + B with A = 0 and B = 6.

In this activity you will see three scatter plots, one at a time. You are to change the A and B values for Y1 = AX + B until you find a function $y = mx + b$ that is a perfect match for the given plot.

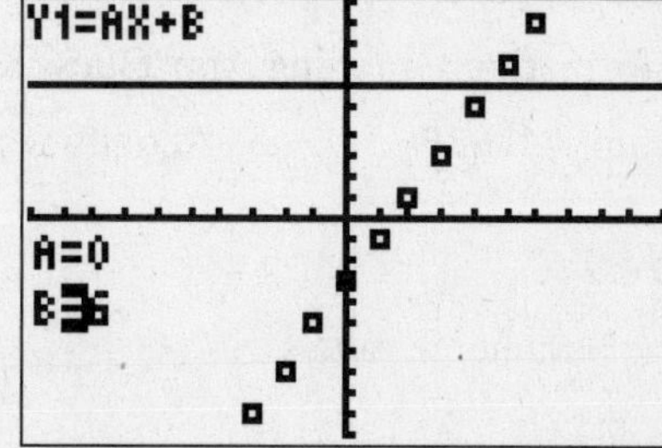

1. Run the Transformation Graphing App and A1L62B. Find a linear function whose graph matches the plot.

2. Switch from Plot1 to Plot2. Press GRAPH to see A1L62B for the second plot. Find a matching linear function.

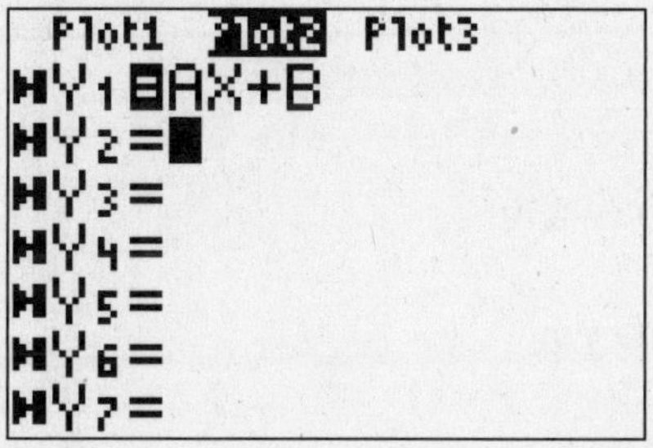

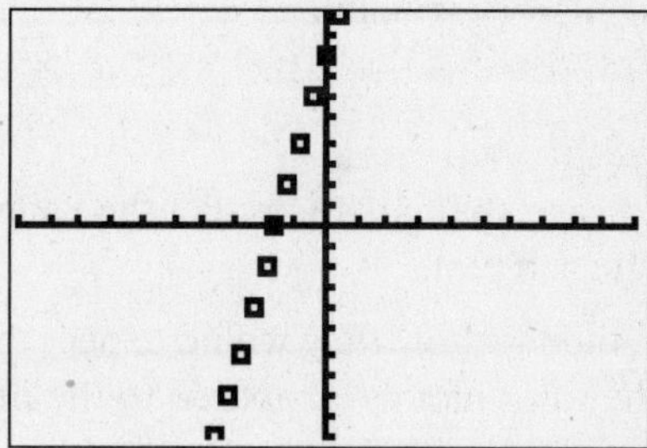

3. Switch from Plot2 to Plot3. Press GRAPH to see A1L62B for the third plot. Find a matching linear function.

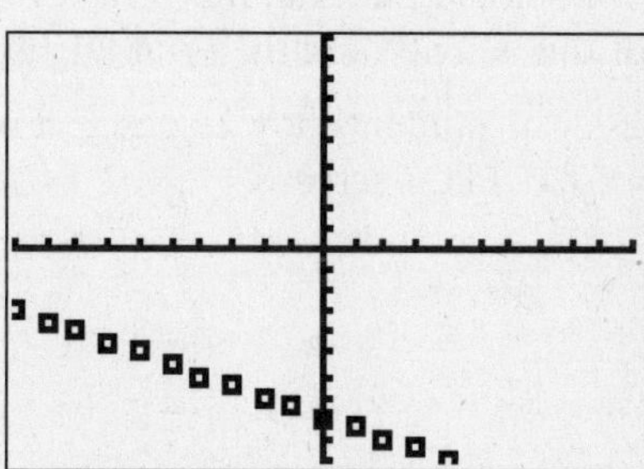

4. Run the Guess My Coefficients (GuesCoef) App. Play the y = mx + b version of the LINEAR game.

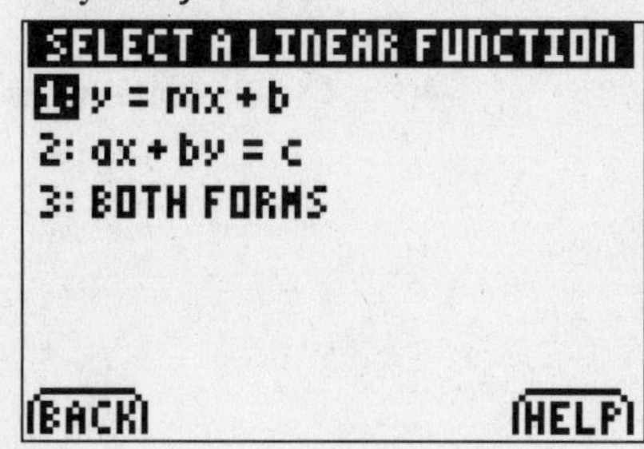

Plot1 Plot2 Plot3
Y1 = AX+B
Y2 =
Y3 =
Y4 =
Y5 =
Y6 =
Y7 =

Extension

Use Plot1 in the Transformation Graphing App. For each equation below, replace data lists L1 and L2 with three ordered pairs that are solutions of the equation. Then find a linear function $y = mx + b$ whose graph contains the three points.

5. $2x + 3y = 6$ **6.** $4x - 2y = 4$ **7.** $x + 3y = 9$

Linear Function Match

Teacher Notes

Activity Objective

Students use the Transformation Graphing App to learn how changing the values of m and b affects the graph of $y = mx + b$.

Time

- 10–15 minutes

Materials/Software

- Transformation Graphing App, Guess My Coefficients App
- Program: A1L62B
- Activity worksheet

Skills Needed

- change parameter values

Classroom Management

- Students can work individually or in pairs depending on the number of calculators available.

Notes

- Encourage students to predict the values of A and B before trying to match each plot.
- Ask students how they would expect the value of A to be different for the third plot as compared to the first two plots.
- Students should uninstall the Transformation Graphing App at the end of this activity, and then run DEFAULT to reset their calculators.
- Discuss with students how to deselect and select a plot in either the Y= or STAT PLOT screens.

Answers

1. $y = 2x - 3$ **2.** $y = 5x + 8$ **3.** $y = -0.5x - 8$ **4.** Check students' work.

5–7. Ordered pairs may vary: Samples are given.

5. $(0, 2), (3, 0), (6, -2); y = -0.67x + 2$, or $y = -\frac{2}{3}x + 2$

6. $(0, -2), (1, 0), (2, 2); y = 2x - 2$

7. $(0, 3), (9, 0), (3, 2); y = -0.33x + 3$, or $y = -\frac{1}{3}x + 3$

Name ____________________ Class ____________ Date ____________

Line 'Em Up

Activity 4

FILES NEEDED: Transformation Graphing App
Program: A1L62C

The A1L62C startup screen shows 14 points plotted in Quadrant I. It also shows the graph of a linear function in the form Y1 = AX + B with A = 1 and B = −1.

In this activity you are to change the values of A and B to find equations for the lines described. Write the equation of each line you find.

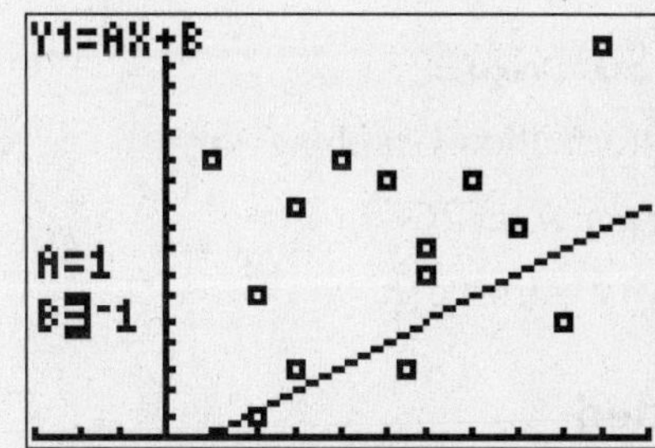

1. Run A1L62C. Write the equation of the line shown on the screen. It passes through one of the plotted points. What are the coordinates of that point? Check that it is a solution of your equation.

2. Change the values of A and B and find a line that passes through two points. Make a guess as to how many such lines you could find.

3. Find a line that passes through three plotted points. Can you find any that pass through four plotted points?

4. Find horizontal lines that pass through at least two points each.

5. Find vertical lines that pass through at least two points each. (*Hint:* You cannot do this by changing the values of A and B.)

6. Can you find a line with positive slope that has on each side of it approximately one half of the points that it doesn't pass through? Can you find a line with negative slope that separates the points in the same way?

Extension

7. If no three of the 14 points were to be collinear, about how many different equations would have graphs passing through two of the 14 points? (Compare your answer with your guess in Question 2.)

8. If there would be only one set of three collinear points among the 14 given points, then how many different equations would have graphs passing through at least two points each?

9. If there was exactly one set of four collinear points and only one (different) set of three collinear points, then how many different equations would have graphs passing through at least two points each?

Line 'Em Up

Teacher Notes

Activity Objective

Students use the Transformation Graphing App to learn how changing the values of m and b affects the graph of $y = mx + b$.

Time

- 20–25 minutes

Materials/Software

- Transformation Graphing App
- Program: A1L62C
- Activity worksheet

Skills Needed

- change parameter values

Classroom Management

- Students can work individually or in pairs depending on the number of calculators available.

Notes

- Sometimes a line will darken the interior of a point mark. Point out that a dark interior does not necessarily mean the point is on the line, nor does a light interior mean that a point is not on a line.
- Students can use TRACE to find the coordinates of a plotted point or a point on a line.

Answers

1. $y = x - 1$; (2, 1)

2. Answers may vary. (See Question 7.)

3. Answers may vary. Sample: The graph for $y = -x + 13$ passes through three plotted points. No line passes through four plotted points.

4. $y = 3, y = 11, y = 12$

5. $x = 2, x = 3, x = 6$

6. Check students' work.

7. 91 equations **8.** 89 equations **9.** 84 equations

Name ______________________ Class ____________ Date ____________

Writing Equations of Lines

Activity 5

FILES NEEDED: Guess My Coefficients App

In this activity you practice writing equations of lines in both slope-intercept form $y = mx + b$ and standard form $ax + by = c$.

Write a description of the graph of each equation.

1. $y = 3x + 5$ **2.** $y = -2x - 1$ **3.** $y = 7$ **4.** $x = -2$

Write an equation for each graph shown below. Use $y = mx + b$ form. The scale on each axis is marked in unit intervals.

5.

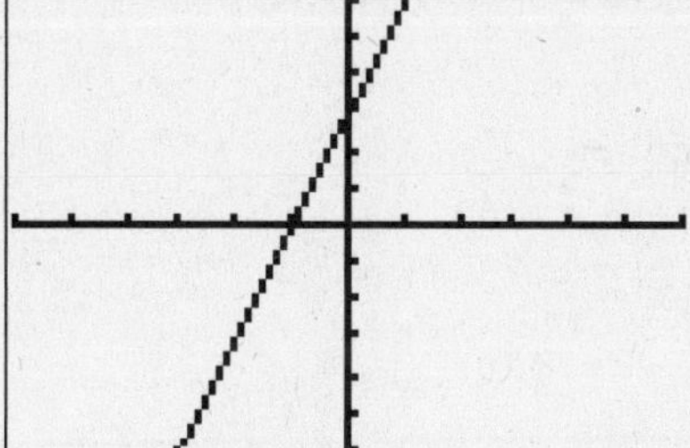

6.

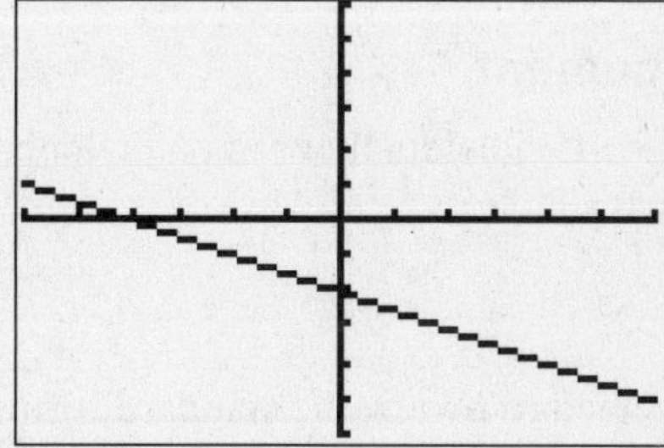

Find the x- and y-intercepts for each line given below. On graph paper, use the intercepts to locate two points of the graph. Then draw the graph.

7. $2x + 3y = 6$ **8.** $x - 2y = 4$ **9.** $4x + 2y = 4$ **10.** $-2x - 3y = 6$

Write an equation for each graph given below. Use $ax + by = c$ form.

11.

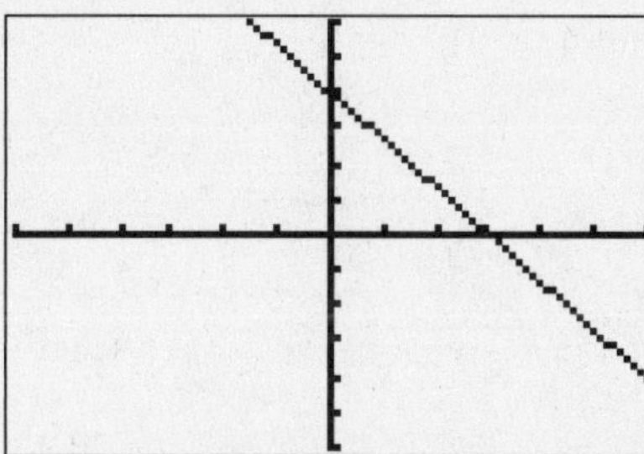

12.

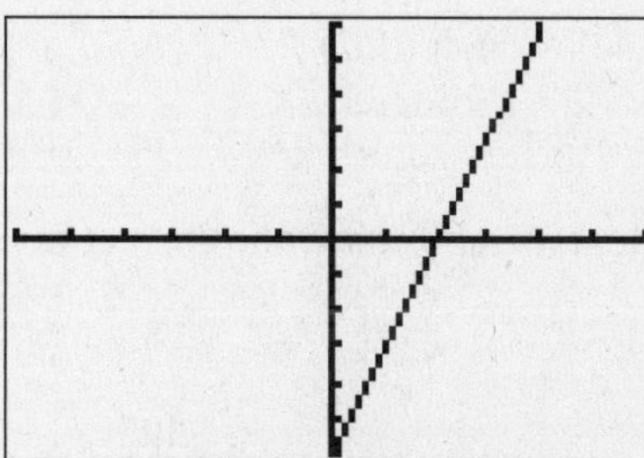

13. For more practice, run the Guess My Coefficients (GuesCoef) App. Play the BOTH FORMS version of the LINEAR game.

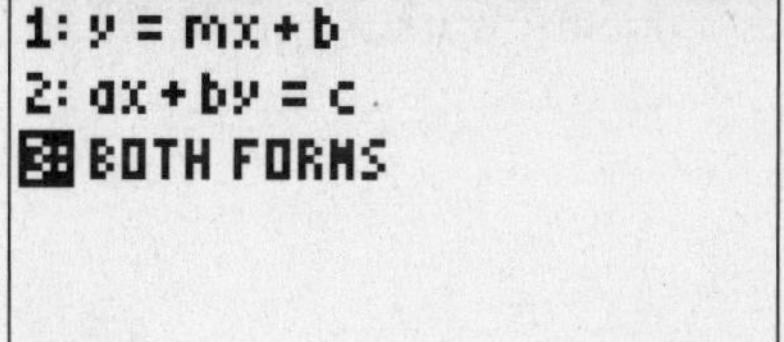

Writing Equations of Lines

Teacher Notes

Activity Objective

Students use the Guess My Coefficients App to practice writing equations of lines in slope-intercept form, $y = mx + b$, and standard form, $ax + by = c$.

Time

- 15–25 minutes

Materials/Software

- Guess My Coefficients App
- Activity worksheet

Classroom Management

- Students can work individually or in pairs depending on the number of calculators available.

Notes

- Students may need to review the standard form of an equation.

Answers

1–4. Answers may vary. Samples are given.

1. a line with slope 3 and y-intercept 5

2. a line with slope –2 and y-intercept –1

3. a horizontal line through (0, 7)

4. a vertical line through (–2, 0)

5. $y = 3x + 3$

6. $y = -\frac{1}{2}x - 2$

7-10. Check that students' graphs have the correct intercepts.

7. $(3, 0), (0, 2)$

8. $(4, 0), (0, -2)$

9. $(1, 0), (0, 2)$

10. $(-3, 0), (0, -2)$

11. $4x + 3y = 12$

12. $6x - 2y = 12$

13. Check students' work.

Name ______________________ Class ______________ Date ______________

Trend Lines

Activity 6

FILES NEEDED: Transformation Graphing App
Programs: A1L66A, A1L66B, A1L66C

Each program for this activity graphs the linear equation $y = mx + b$ as Y1 = AX + B.

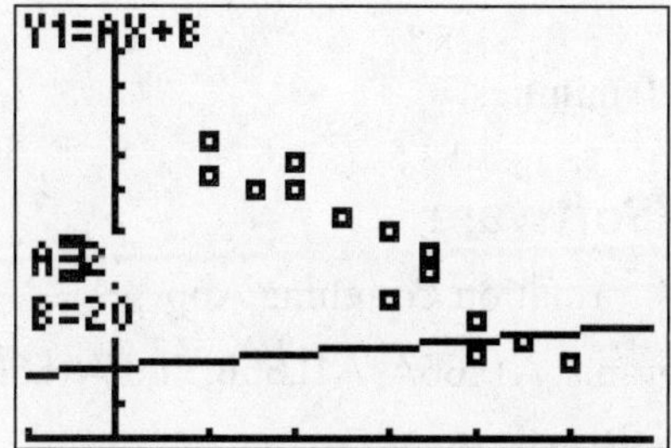

In this activity, you will use $y = mx + b$ to explore trend lines.

If a scatter plot suggests a relationship between two data sets, a *trend line* provides a mathematical model for the relationship. For a line to be a trend line, you want the points of the scatter plot to cluster around the line and the points of the line to approximate the points of the scatter plot.

Each program for this activity relates to an exercise in Lesson 6-6 of your textbook, pages 320–321. Run the program. Manipulate the line shown in the window to make the best trend line you can. Write the equation of the line. Then answer the related question(s).

1. Program A1L66A (Exercise 4) — The scatter plot compares study time x to response speed y in a memory test. Use your trend line. What is an estimated response speed for a study time of 5.5 minutes?

2. Program A1L66B (Exercise 7) — The scatter plot compares latitude (°N) with average temperature (°F). Does your trend line have positive slope or negative slope? What does the sign of the slope tell you about the relationship between latitude and temperature? At what latitude would you expect to find an average temperature of 45°F?

3. Program A1L66C (Exercise 11) — The scatter plot compares air temperature and wind-chill temperature when the wind speed is 15 mi/h. Use your trend line. Estimate the wind-chill temperature when the air temperature is 23°F.

Extension

Work with another student. Compare your trend lines.

4. For the memory test (Question 1 above), explain any differences you find in your trend lines. Compare your explanations.

5. You will likely find that your wind-chill-temperature trend lines (Question 3) are more alike than those for the memory test and the temperatures in northern latitudes. Why might this be so?

Trend Lines

Teacher Notes

Activity Objective

Students use the Transformation Graphing App to make trend lines.

Time

- 15–20 minutes

Materials/Software

- Transformation Graphing App
- Programs: A1L66A, A1L66B, and A1L66C
- Activity worksheet

Skills Needed

- start an App
- change parameter values

Classroom Management

- Students can work individually or in pairs depending on the number of calculators available.
- Use TI Connect™ software, TI-GRAPH LINK™ software, the TI-Navigator™ system, or unit-to-unit links to transfer A1L66A, A1L66B, and A1L66C to each calculator.

Notes

- When the equal sign next to the parameter is highlighted, students can change parameters A or B by entering the values directly.
- Use value in the CALC menu to find a y-value for a given x-value on the trend line.

Answers

Answer may vary. Samples are given.

1. $y = -16x + 98$; 10 s
2. $y = -0.9x + 87$; negative; As the latitude increases, the average temperature decreases. ≈ 46°N
3. $y = 1.5x - 34$; 0.5°F
4. Check students' work.
5. The data are very close to being linear.

Name ______________________ Class ______________ Date ______________

Absolute Value Graphs

Activity 7

FILES NEEDED: Transformation Graphing App
Program: A1L67

This activity uses the Transformation Graphing App to model bank shots on a pool table. A bank shot bounces off a cushion (the inner rim of the pool table top) at the same angle that it approaches the cushion.

Rules and Scoring

In A1L67, the 11 plotted points represent pool balls. The graph is that of an absolute value equation of the form Y1 = A abs(X – B) + C.

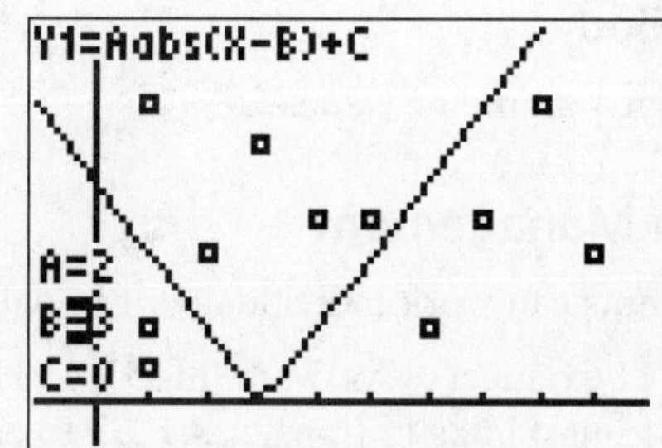

Your task is to find values for A, B, and C so that the graph strikes the pool balls. You can hit a pool ball with any part of the graph. The vertex of the graph must sit on the *x*-axis in the first quadrant.

Score one point for each ball struck.

EXAMPLE: The score for the startup graph in A1L67 is 0 because the graph hits no pool balls.

OBJECT: Find values for A, B, and C to score the greatest number of points.

1. Do the rules limit the values of C you can use? Explain.
2. How does changing the value of A affect the shape of the graph?
3. How does the object of the game limit the values of A you can use?
4. How does changing the value of B affect the graph?
5. Finding a graph whose score is 1 is easy. Why?
6. Finding a graph whose score is 2 is easy. Why?
7. For what scores greater than 2 can you find a graph? Write the equation for your highest-scoring graph. What is its score?

Extension

8. Explain how to modify the game so that the graph uses the top of the screen as the cushion for its bank shots.
9. Explain why it would be difficult to use the left or right side of the screen as the cushion.

Absolute Value Graphs

Teacher Notes

Activity Objective

Students use the Transformation Graphing App to explore absolute value graphs.

Time

- 30–45 minutes

Materials/Software

- Transformation Graphing App
- Program: A1L67
- Activity worksheet

Skills Needed

- change parameter values

Classroom Management

- Students can work individually or in pairs depending on the number of calculators available.
- Use TI Connect™ software, TI-GRAPH LINK™ software, the TI-Navigator™ system, or unit-to-unit links to transfer A1L67 to each calculator.

Notes

- The text discusses the forms $y = |x| + k$ and $y = |x - h|$, but does not deal directly with $y = a|x - h| + k$.
- Investigating the parameters A, B, and C separately may be helpful for students who have trouble seeing how each parameter affects the graph.

Answers

1. Because the vertex must be on the x-axis, the minimum value must be $y = 0$, so C must equal 0.

2. The value of A affects how narrow or wide the V-shape is.

3. To hit any points, A must by greater than zero.

4. Changing B translates the graph to the left or right.

5. There are infinitely many lines through any point.

6. Two points determine a line. You can pass a branch of the graph through any two points.

7. Check students' work.

8. Let C = 11. Use negative values for A.

9. Y1 gives the graph of a function. A function does not have two points of its graph in line vertically.

Name ____________________ Class ____________ Date ____________

Tortoise and the Hare

Activity 8

FILES NEEDED: Transformation Graphing App
Program: A1L71

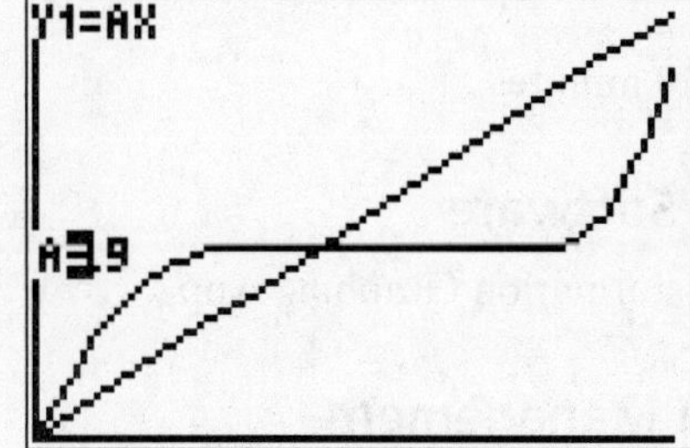

The diagram at the right models the race of the tortoise and the hare. To understand the model you have to first answer Questions 1 and 2 correctly. Their answers are shown upside down at the bottom of this page.

1. Run A1L71. Which graph belongs to each animal? Explain.
2. The horizontal axis represents time. Does the vertical axis represent distance or speed? Explain.
3. For the correct answer to Question 2, explain what each graph shows. Also, for the diagram above, explain what the top of the screen represents and how the diagram shows that the turtle won the race.

The distance run by the turtle is modeled by the equation $y = ax$, graphed above as Y1 = AX. You can vary the slope, a, of the line and determine values of a for which the turtle would win, lose, or tie.

When you use your calculator for the remaining questions, change A values by 0.1. You should enter directly each new value that you want to view.

4. For what values of A will the tortoise win the race?
5. For what values of A will the tortoise lose the race?
6. For what value of A will the tortoise and the hare tie in the race?
7. For what values of A will the tortoise not only lose the race but never be ahead of the hare?
8. For approximately what value of A will the tortoise pass the hare just as soon as the hare stops to take a rest?
9. For what values of A will the tortoise be ahead for part of the race but still lose?
10. For the model, what does a represent? How does a relate to the moral of the fable, "Slow and steady wins the race"?
11. On the diagram above, draw an interval along the time axis over which you think the turtle was running faster than the hare. Explain your drawing.

1. The hare took a nap during the race. The graph with the flat part belongs to the hare. The graph that's a line belongs to the tortoise.
2. Distance; if the vertical axis showed speed, the turtle's graph would be a horizontal line.

Tortoise and the Hare

Teacher Notes

Activity Objective

Students use the Transformation Graphing App to interpret slopes and graphs.

Time

- 15–20 minutes

Materials/Software

- Transformation Graphing App
- Program: A1L71
- Activity worksheet

Classroom Management

- Students can work individually or in pairs depending on the number of calculators available.
- Use TI Connect™ software, TI-GRAPH LINK™ software, the TI-Navigator™ system, or unit-to-unit links to transfer A1L71 to each calculator.

Notes

- Discuss with the class the correct interpretation of the graphs, including where the finish line is located on the screen.
- Remind students that they can enter values directly into A=. Type the value; then press ENTER.
- Remind students to uninstall the Transformation Graphing App when they complete the activity, and then run DEFAULT.

Answers

3. The line shows the tortoise's steady progress. The other graph shows the hare running quickly, covering distance faster than the tortoise, but then slowing down and coming to a stop for a nap. The tortoise passes the hare at the point of intersection. Later the hare starts moving again very fast, but the tortoise reaches the finish line (the top of the screen) first.

4. $A \geq 0.9$

5. $A \leq 0.8$

6. some value between 0.8 and 0.9

7. $A < 0.5$

8. $A \approx 1.5$

9. $0.5 < A < 0.8$

10. The tortoise's speed; the tortoise maintained a steady speed.

11. Check students' work; the interval along which the slope of the hare's graph is less than the slope of the tortoise's graph.

Name ______________________________ Class ________________ Date ________________

Linear Systems I

Activity 9

FILES NEEDED: Transformation Graphing App
Program: A1L72A

You run two battery-powered cars in opposite directions across the classroom floor. You use a motion detector to plot the distance/time graphs shown here.

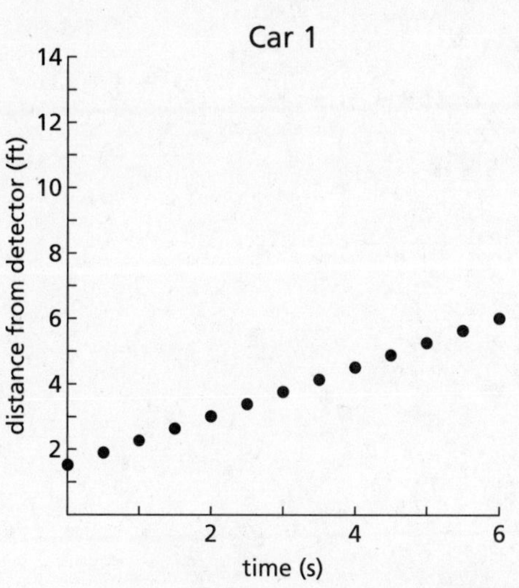

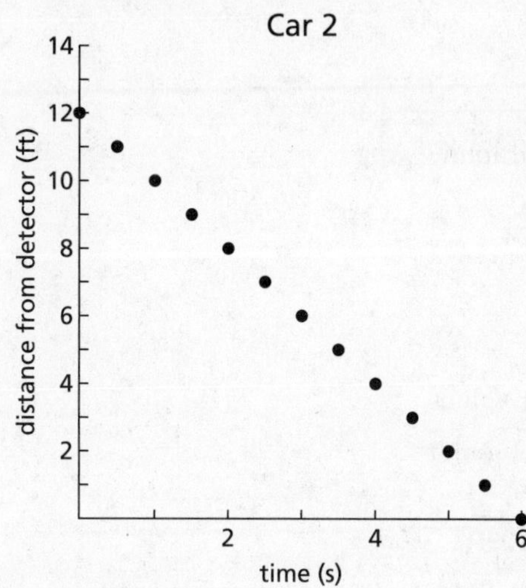

1. One car moved away from the motion detector. The other moved toward it. Which car moved away from the motion detector?

2. About how far from the motion detector did each car start?

3. If both cars started at the same time, when did they pass each other?

4. When they passed, how far was each from the motion detector?

5. Which car moved faster? Explain.

6. Run A1L72A to see the plot for Car 1. The screen also shows the line $y = mx + b$ graphed as Y1 = AX + B with A = 0 and B = 6. Change the values of A and B to find a reasonable model for the distance/ time plot of Car 1. Find values for A and B to 2 decimal places. Write your model in the form $y = mx + b$.

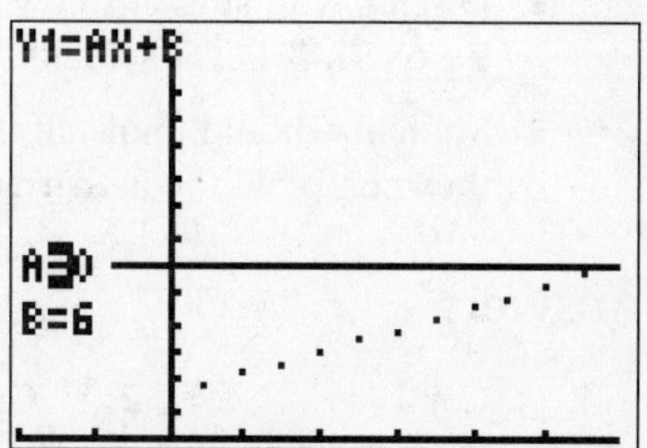

7. Switch from Plot1 to Plot2. Press GRAPH to see A1L72A for Car 2. Use the procedure from Question 6 and find a model for the distance/time plot of Car 2.

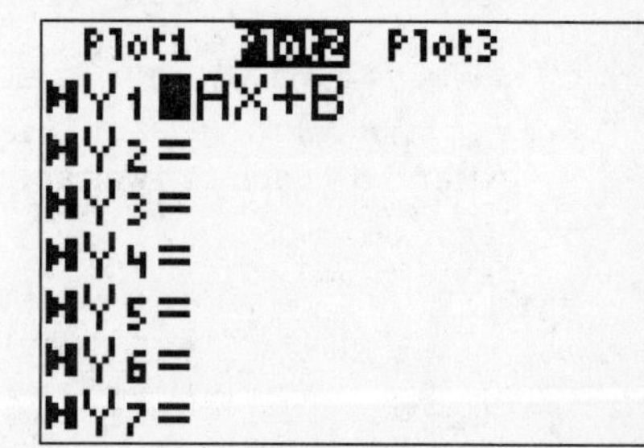

8. Use your models from Questions 6 and 7 and a method of your choice to answer Questions 3 and 4. Check the reasonableness of your answers with the plots at the top of this page.

Linear Systems I

Teacher Notes

Activity Objective

Students use the Transformation Graphing App to interpret graphs and the point of intersection of two lines.

Time

- 10–15 minutes

Materials/Software

- Transformation Graphing App
- Program: A1L72A
- Activity worksheet

Skills Needed

- change parameter values
- select and deselect a plot

Classroom Management

- Students can work individually or in pairs depending on the number of calculators available.
- Use TI Connect™ software, TI-GRAPH LINK™ software, the TI-Navigator™ system, or unit-to-unit links to transfer A1L72A to each calculator.

Notes

- Remind students that they can enter values directly into A= or B=. Type the value; then press ENTER.
- Discuss with students how to deselect and select a plot in either the Y= or STAT PLOT screens.
- Students should uninstall the Transformation Graphing App when they complete the activity, and then run DEFAULT.

Answers

1. Car 1 **2.** Car 1: ≈1.5 ft; Car 2: 12 ft **3.** about 4 s

4. about 4 ft **5.** Car 2; It traveled 12 ft in 6 s, while car 1 traveled about 4.5 ft in 6 s.

6. $y = 0.75x + 1.50$ **7.** $y = -2x + 12$

8. After 3.8 s each car was about 4.4 ft from the detector.

Name ______________________ Class ____________ Date ____________

Linear Systems II

Activity 10

FILES NEEDED: Transformation Graphing App
Program: A1L72B

You race two battery-powered cars in the same direction across the classroom floor. You use a motion detector to plot the distance/time graphs shown here.

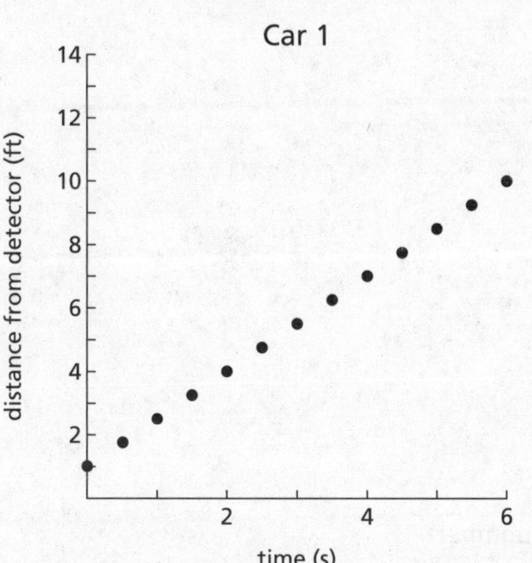

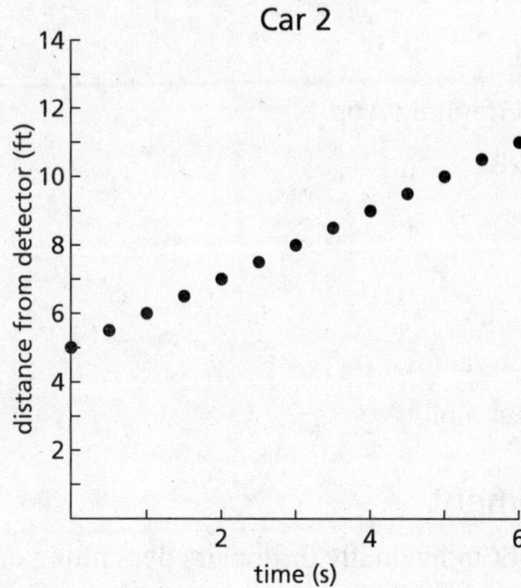

1. About how far from the motion detector did each car start?

2. How much of a head start did you give Car 2?

3. Which car moved faster? Explain.

4. If both cars started at the same time and the race lasted exactly 6 s as shown in the plots, did Car 1 catch Car 2? If so, how far were the cars from the motion detector?

5. Run A1L72B to see the plot for Car 1. The screen also shows the line $y = mx + b$ graphed as Y1 = AX + B with A = 0 and B = 6. Change the values of A and B to find a reasonable model for the distance/time plot of Car 1. Write your model in the form $y = mx + b$.

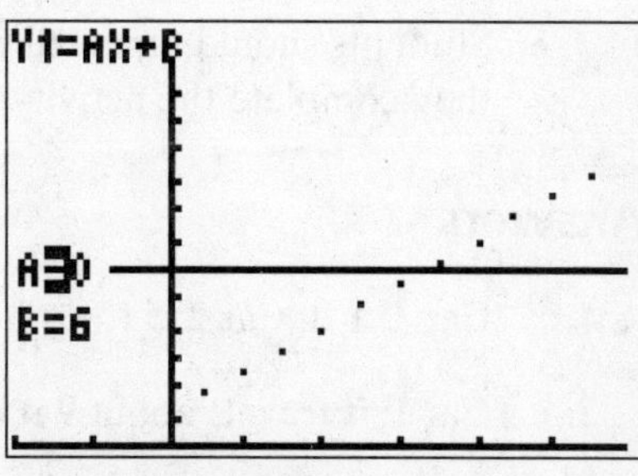

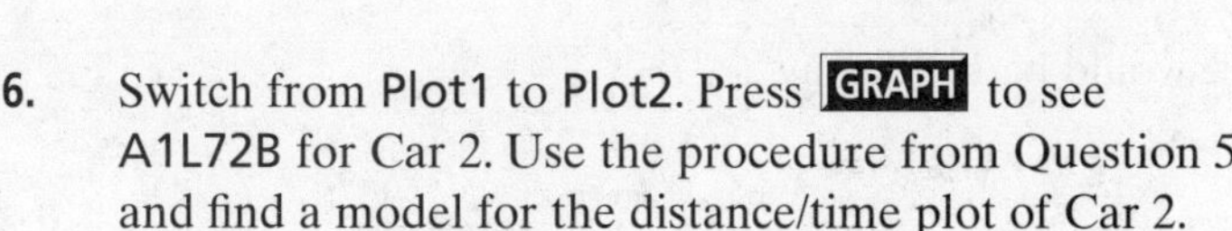

6. Switch from Plot1 to Plot2. Press GRAPH to see A1L72B for Car 2. Use the procedure from Question 5 and find a model for the distance/time plot of Car 2.

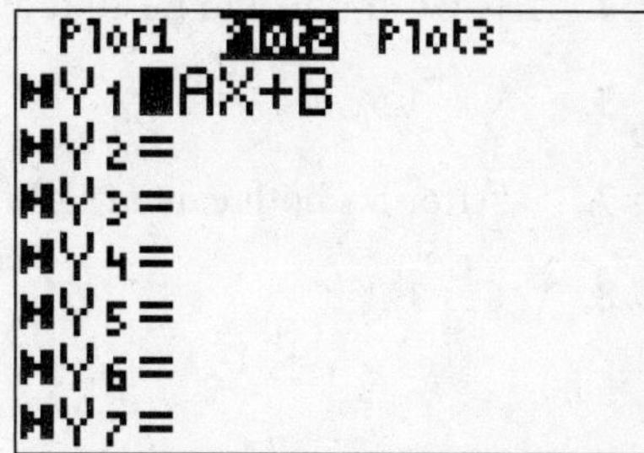

7. Use your models from Questions 5 and 6 and a method of your choice to find when and where Car 1 catches Car 2. Check the reasonableness of your answers with the plots above.

8. If the race lasted 10 s, which car would be farther from the motion detector? How much farther?

Linear Systems II

Teacher Notes

Activity Objective

Students use the Transformation Graphing App to develop interpretations of linear data.

Time

- 10–15 minutes

Materials/Software

- Transformation Graphing App
- Programs: A1L72B
- Activity worksheet

Skills Needed

- change parameter values
- select and deselect a plot

Classroom Management

- Students can work individually or in pairs depending on the number of calculators available.
- Use TI Connect™ software, TI-GRAPH LINK™ software, the TI-Navigator™ system, or unit-to-unit links to transfer A1L72B to each calculator.

Notes

- Remind students that they can enter values directly into A= or B=. Type the value; then press ENTER.
- Students should uninstall the Transformation Graphing App when they complete the activity, and then run DEFAULT.

Answers

1. Car 1: 1 ft; Car 2: 5 ft

2. 4 feet

3. Car 1; it travels about 9 ft in the same time that Car 2 travels about 6 ft.

4. No, Car 1 would be 10 ft away while Car 2 would be 11 ft away.

5. $y = 1.5x + 1$

6. $y = x + 5$

7. After 8 s both cars are 13 ft from the motion detector.

8. Car 1; 1 ft

Name ______________________ Class ______________ Date ______________

Linear Inequality Graphs

Activity 11

FILES NEEDED: Inequality Graphing App

In this activity you practice graphing linear inequalities. After you graph each inequality by hand, use the Inequality Graphing App to check your work.

Follow this example to graph $y > x + 1$ using the Inequality Graphing App Inequal.

In the Y1= line of the Y= screen, select "=." In the list of inequality signs at the screen bottom, note that ">" is above F4. Press ALPHA F4.

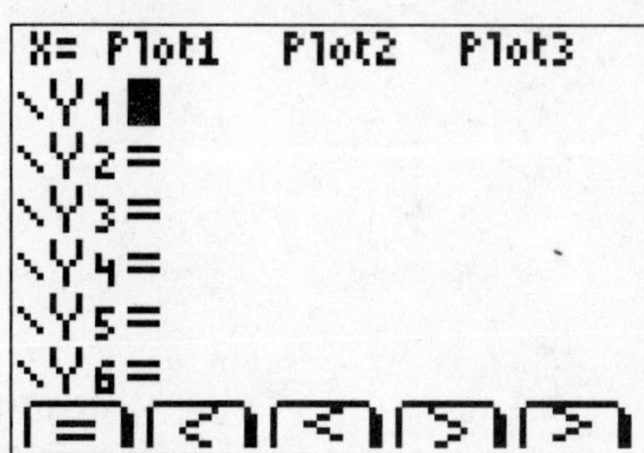

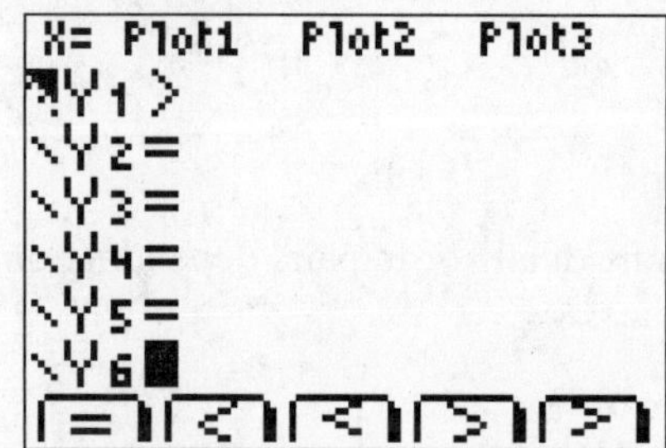

Enter Y1 > X + 1 and . . .

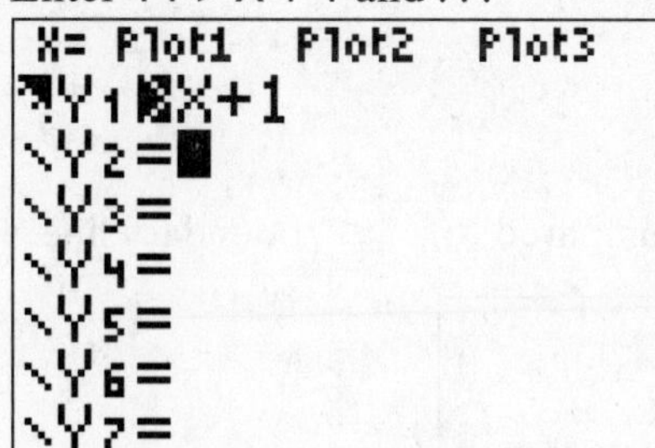

press GRAPH.

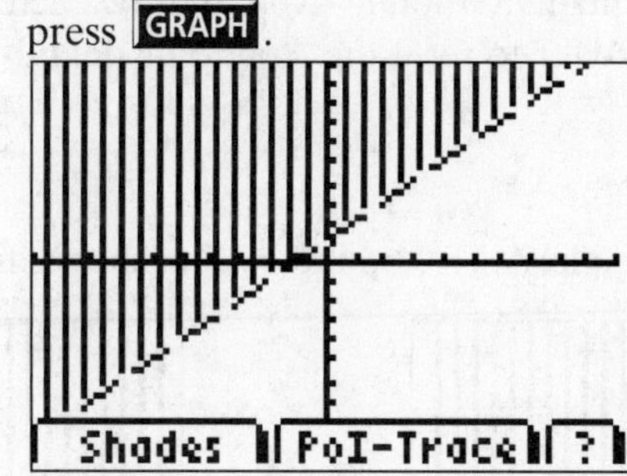

On graph paper, graph each inequality given below. Then use the Inequality Graphing App to draw each graph. Compare graphs. Describe how the screen graph distinguishes a "<" boundary line from a "≤" boundary line.

1. $y < 2x + 5$ **2.** $y \leq 7 - 3x$ **3.** $y \geq 3x - 6$ **4.** $6x + 2y > 10$

Extension

5. Use the Inequality Graphing App to graph Y1 ≤ −3X + 4 and Y2 > 5X − 2 on one screen. Press ALPHA F2 and experiment with the SHADES menu.

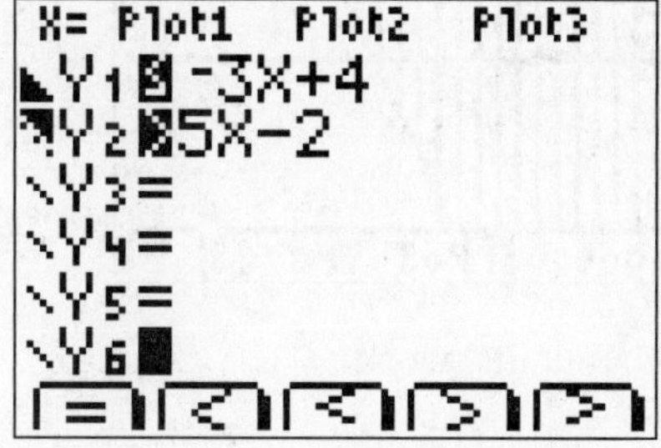

Linear Inequality Graphs

Teacher Notes

Activity Objective

Students use the Inequality Graphing App to practice graphing linear inequalities.

Time

- 15–25 minutes

Materials/Software

- Inequality Graphing App
- Activity worksheet

Classroom Management

- Students can work individually or in pairs depending on the number of calculators available.

Notes

- Before using the Inequality Graphing App to draw each graph, students may wish to run DEFAULT to reset the App to its default values.

Answers

1–4. Answers may vary. Sample: A "$<$" boundary line shows distinct gaps compared to a "$\leq$" boundary line.

1.

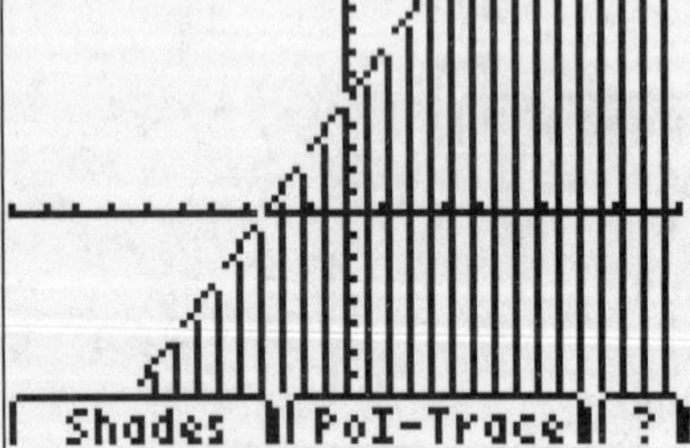

2.

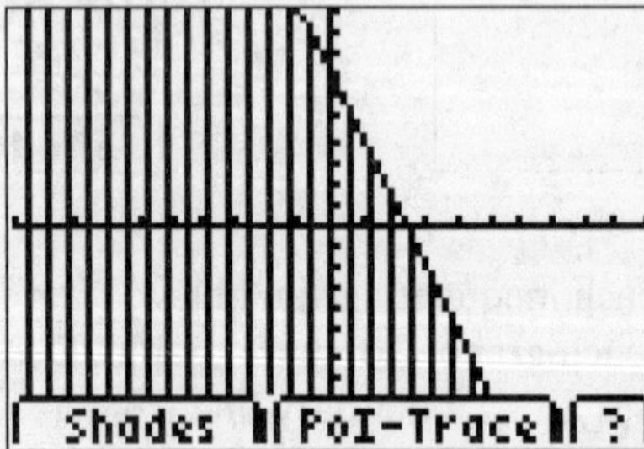

3.

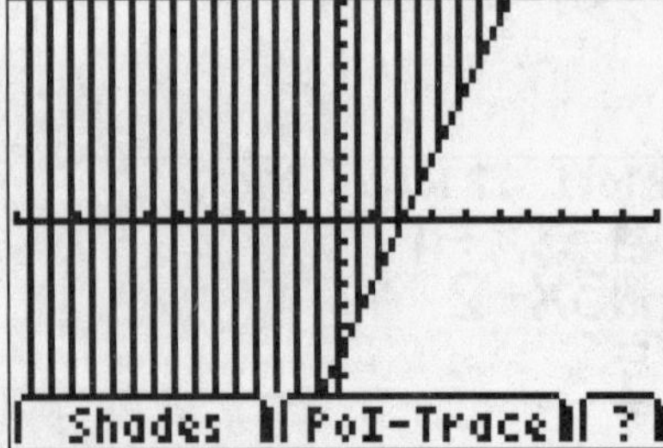

4.

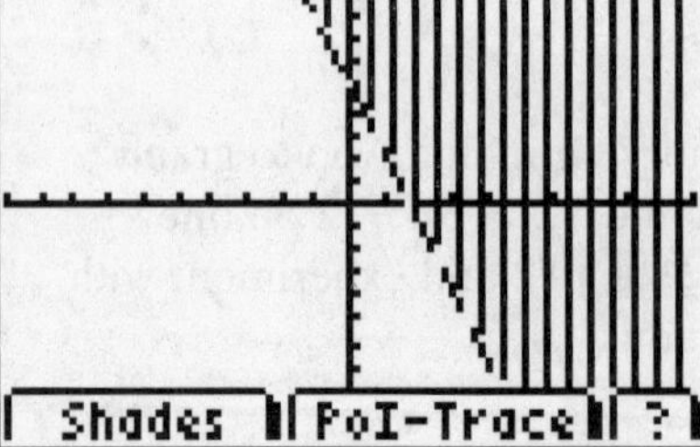

5. Check students' work.

Name ______________________ Class ______________ Date ______________

Linear Inequality Systems

Activity 12

FILES NEEDED: Inequality Graphing App
Program: A1L76

The graphs of two linear inequalities whose boundary lines intersect in one point form four regions in the plane. In this activity you study those four regions.

1. Run A1L76. The screen (below left) shows the graphs of $y = -0.5x + 2$ and $y = 2x - 2$. Which graph belongs to which equation? Explain.

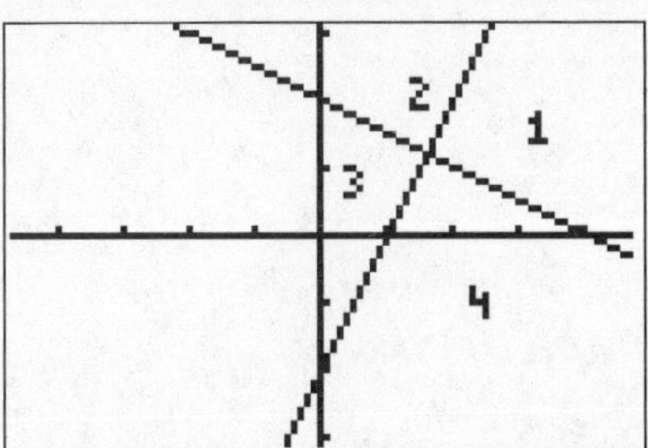

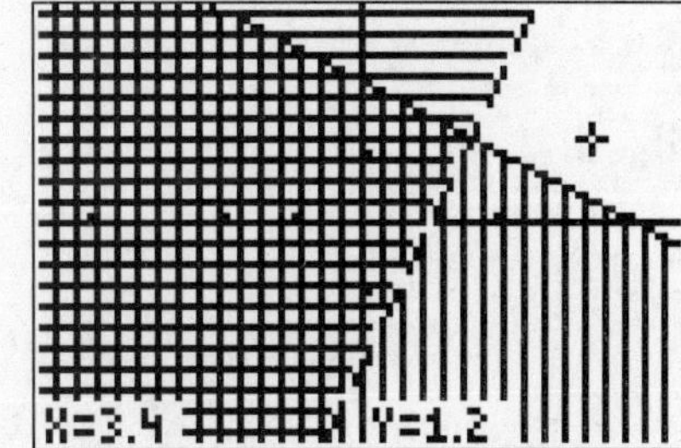

2. Use the Inequality Graphing App. Graph Y1 ≤ −0.5X + 2 and Y2 > 2X − 2 on one screen (above right). Use the cursor to locate two points in each of the four regions of the plane. Record the coordinates in a table like the one below. Complete each line of the table before you locate a new point. You should fill out eight table lines in all.

Region	Number of shadings	Point	Substitute in $y \leq -0.5x + 2$.	Solution?	Substitute in $y > 2x - 2$.	Solution?
1	0	(3.4, 1.2)	$1.2 \leq -0.5(3.4) + 2$ $1.2 \leq 0.3$	no	$1.2 \geq 2(3.4) - 2$ $1.2 \geq 4.8$	no
1						
2						
2						
3						

3. Generalize: Where a point satisfies two inequalities, the graph has __?__ shading(s).
Where a point satisfies only one inequality, the graph has __?__ shading(s).
Where a point satisfies neither inequality, the graph has __?__ shading(s).

Write a system of inequalities for each graph. Check your answer by entering your inequalities as Y1 and Y2.

4.

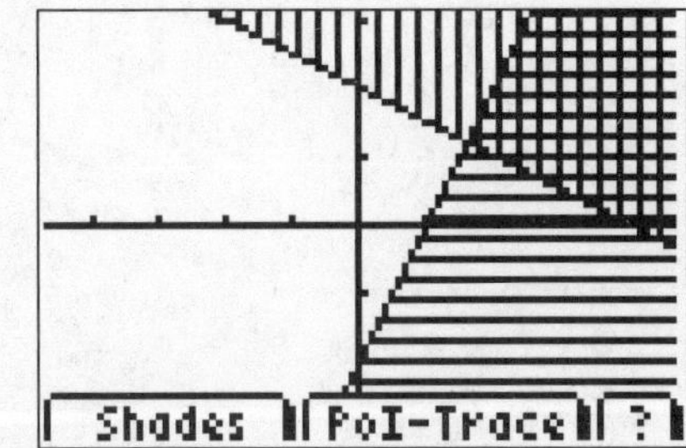

5.

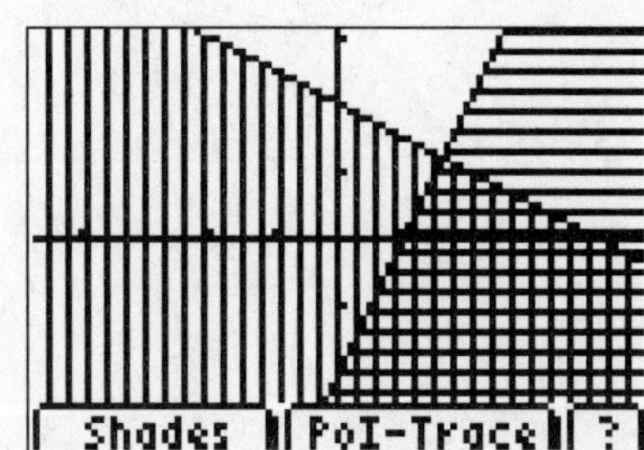

Linear Inequality Systems

Teacher Notes

Activity Objective

Students use the Inequality Graphing App to graph systems of two linear inequalities and to understand the four regions of the plane that result when the boundary lines intersect in one point.

Time

- 20–25 minutes

Materials/Software

- Inequality Graphing App
- Program: A1L76
- Activity worksheet

Skills Needed

- insert inequality sign

Classroom Management

- Students can work individually or in pairs depending on the number of calculators available.

Error Prevention

- Remind students that the "1", "2", "3", and "4" in the startup screen correspond to the shaded regions and not to the quadrants.

Notes

- Use TI Connect™ software, TI-GRAPH LINK™ software, the TI-Navigator™ system, or unit-to-unit links to transfer A1L76 to each calculator.

Answers

1. The line below 1 and 2 is the graph of $y = -0.5x + 2$. The line to right of 2 and 3 is the graph of $y = 2x - 2$. You can tell this by the slopes.
2. Answers may vary. Check students' work.
3. 2; 1; 0
4. $y \geq -0.5x + 2$; $y \leq 2x - 2$
5. $y \leq -0.5x + 2$; $y \leq 2x - 2$

Name ______________________ Class ______________ Date ______________

Exponential Graphs

Activity 13

FILES NEEDED: Transformation Graphing App
Program: A1L87

A1L87 graphs the exponential function $y = ab^x$ as Y1 = AB^X.

In this activity, you will explore the graph of $y = ab^x$.

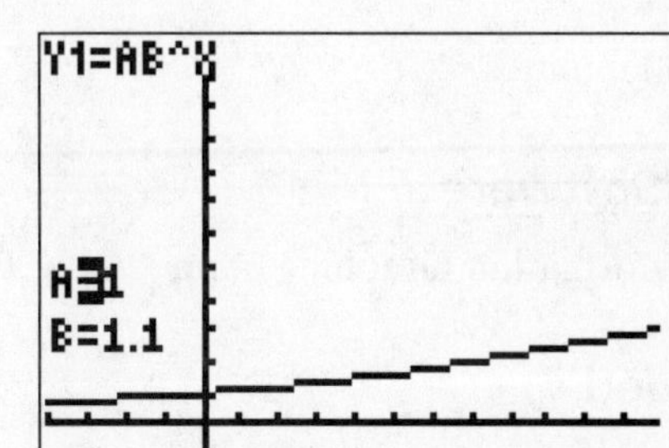

1. Run A1L87. Increment the value of A in steps of 1. (See Animation Option below.) What happens to the graph as A increases?

2. Reset the value of A to 1. Select B and enter values 1.5, 1.75, 2, and 3. What happens to the graph as B increases?

3. Does y have a maximum value when A > 0 and B > 1?

4. Set A $= 8$ and B $= 1$. Describe the graph. Explain why the graph makes sense.

5. Select B and enter values 1, 0.8, 0.6, 0.4, and 0.2. What happens to the graph in Quadrant I as B decreases, $0 <$ B < 1?

6. Fill in the chart. Summarize what you have learned about the effect of B for A > 0.

Describe the graph when B $= 1$.	
Describe the graph when $0 <$ B < 1.	
Describe the graph when B > 1.	
What happens to the graph when B increases from 1?	
What happens to the graph when B decreases from 1 (B > 0)?	

Extension

7. For A > 0 and $0 <$ B < 1, what happens to Y as X gets very large?

8. Suppose A < 0. How are the graphs affected?

Animation Option

In Question 1, set Step $= .5$, A $= 1$, B $= 1.1$, and Max $= 5$. Press GRAPH to animate.

Exponential Graphs

Activity Objective

Students use the Transformation Graphing App to explore how values of a and b affect the graph of $y = ab^x$.

Time

- 25–30 minutes

Materials/Software

- Transformation Graphing App
- Program: A1L87
- Activity worksheet

Skills Needed

- change parameter values

Classroom Management

- Students can work individually or in pairs depending on the number of calculators available.
- Circulate around the room as students work to help individuals and groups discover the concepts.

Notes

- Reminder: As noted on p. vi, each Activity page assumes that you activate the appropriate App at the start of the activity.
- Students can enter values for A and B directly by typing the number and pressing ENTER.
- By the end of the activity, discuss A as the y-intercept, a starting place, and B reflecting a rate of growth or decay.
- Students should uninstall the Transformation Graphing App at the end of this activity and then run DEFAULT.

Answers

1. The y-intercept increases. **2.** The graph rises more quickly. **3.** No; as x increases, y increases.

4. The horizontal line $y = 8$. The function becomes $y = 8(1)^x$ or $y = 8$.

5. The graph gets closer and closer to the x-axis.

6. A horizontal line through $y =$ A; a decay function with A as the y-intercept; a growth function with A as the y-intercept; the graph rises more quickly; the graph gets closer and closer to the x-axis.

7. Y gets very close to zero.

8. The graphs reflect across the x-axis.

Name ______________________ Class ______________ Date ______________

Exponential Function Match

Activity 14

FILES NEEDED: Transformation Graphing App
Program: A1L88

A1L88 graphs the exponential function $y = ab^x$ as Y1 = AB^X with A = 1 and B = 1.1.

In this activity you will see three scatter plots, one at a time. You are to change the A and B values for Y1 = AB^X until you find a function $y = ab^x$ that is a perfect match for the given plot.

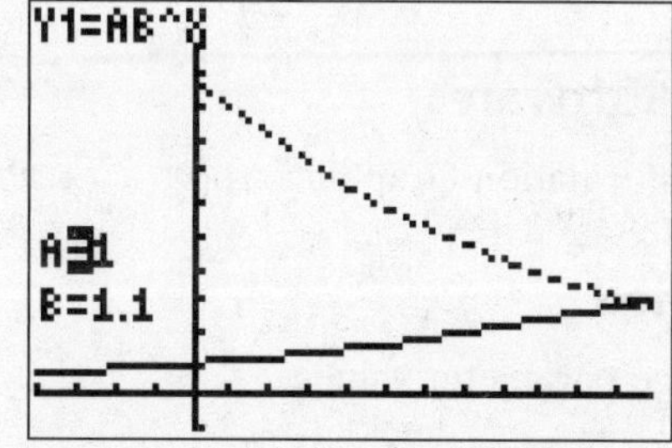

1. Run A1L88. Find an exponential function whose graph matches the plot.

2. Switch from Plot1 to Plot2. Press GRAPH to see A1L88 for the second plot. Find a matching exponential function.

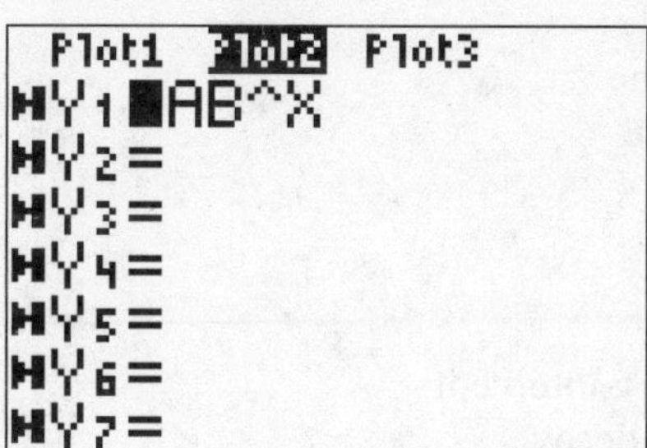

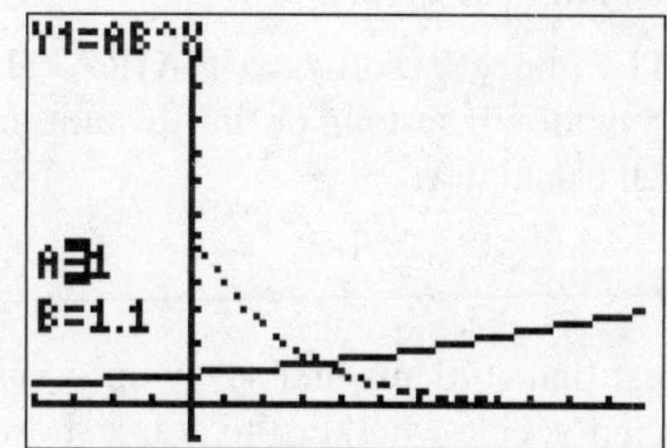

3. Switch from Plot2 to Plot3. Press GRAPH to see A1L88 for the third plot. Find a matching exponential function.

4. By looking at the graphs, how can you tell whether to use values of B with $B > 1$ or with $0 < B < 1$?

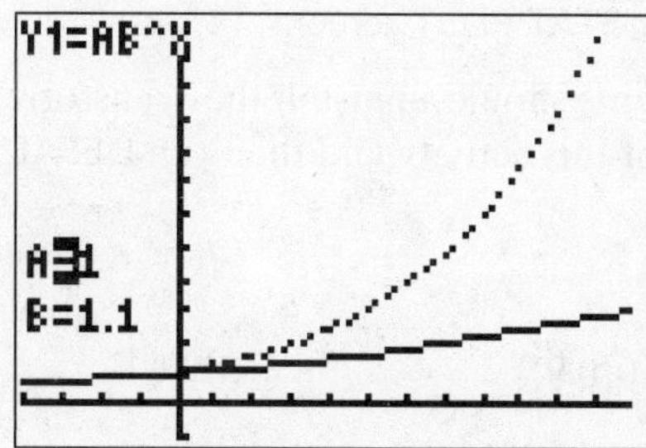

Exponential Function Match

Teacher Notes

Activity Objective

Students use the Transformation Graphing App to learn how changing the values of a and b affects the graph of $y = ab^x$.

Time

- 20–25 minutes

Materials/Software

- Transformation Graphing App
- Program: A1L88
- Activity worksheet

Skills Needed

- change parameter values
- select and deselect a plot

Classroom Management

- Students can work individually or in pairs depending on the number of calculators available.
- Use TI Connect™ software, TI-GRAPH LINK™ software, the TI-Navigator™ system, or unit-to-unit links to transfer A1L88 to each calculator.

Notes

- Suggest that students first select an A value that matches the y-intercept and then decide whether the graph shows growth ($B > 1$) or decay ($B < 1$). Use this information to select a starting value for B.
- Discuss with students how to deselect and select a plot in either the Y= or STAT PLOT screens.
- Students should uninstall the Transformation Graphing App at the end of this activity and then run DEFAULT.

Answers

1. $y = 10(0.9)^x$
2. $y = 6(0.65)^x$
3. $y = 1(1.25)^x$
4. If the graph "grows," or rises to the right, use $B > 1$. If it shows decay, use $0 < B < 1$.

Name ______________________ Class ____________ Date ____________

Multiplying Binomials

Activity 15

FILES NEEDED: Cabri® Jr.
AppVar: A1L93

A1L93 shows the area model for the product of two binomials.

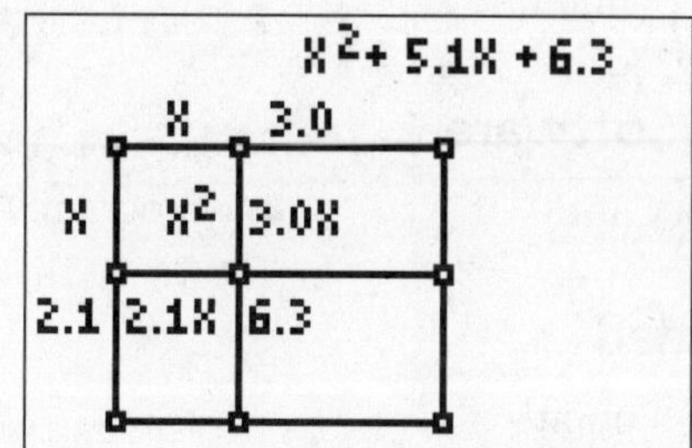

1. Each rectangle in the model has length, width, and area. For the first diagram shown in A1L93 (at the right), list the length, width, and area of each of the five rectangles below.

Rectangle	Length	Width	Area
top left			
top right			
bottom left			
bottom right			
large rectangle			

2. You can change the dimensions of the rectangle by dragging the top-right vertex and the lower-left vertex. Change both dimensions and complete the table below for your new model.

Rectangle	Length	Width	Area
top left			
top right			
bottom left			
bottom right			
large rectangle			

3. Predict the values that complete the table below. Then test your predictions.

Rectangle	Length	Width	Area
top left	x	x	
top right	4		
bottom left		2	
bottom right			
large rectangle			

Extension

4. Use pencil and paper. Generalize the model by drawing a diagram and making a table for which the top-right rectangle has length a and the bottom left rectangle has width b.

Multiplying Binomials

Teacher Notes

Activity Objective

Students use Cabri® Jr. to develop understanding of how to multiply binomials.

Time

- 15–20 minutes

Materials/Software

- App: Cabri® Jr.
- Program: A1L93
- Activity worksheet

Skills Needed

- drag a point

Classroom Management

- Students can work individually or in pairs depending on the number of calculators available.
- Use TI Connect™ software, TI-GRAPH LINK™ software, the TI-Navigator™ system, or unit-to-unit links to transfer A1L93 to each calculator.

Notes

- Remind students that Cabri® Jr. rounds values to the nearest tenth.

Answers

1.

Rectangle	Length	Width	Area
top left	x	x	x^2
top right	3.0	x	$3.0x$
bottom left	x	2.1	$2.1x$
bottom right	3.0	2.1	6.3
large rectangle	$x + 3.0$	$x + 2.1$	$x^2 + 5.1x + 6.3$

2. Check students' work.

3.

Rectangle	Length	Width	Area
top left	x	x	x^2
top right	4	x	$4x$
bottom left	x	2	$2x$
bottom right	4	2	8
large rectangle	$x + 4$	$x + 2$	$x^2 + 6x + 8$

4. The large rectangle should have length $x + a$, width $x + b$, and area $(x + a)(x + b) = x^2 + (a + b)x + ab$.

Name ______________________ Class ______________ Date ______________

Square of a Binomial

Activity 16

FILES NEEDED: Cabri® Jr.
AppVar: A1L94

A1L94 shows the area model for the square of a binomial.

1. Each rectangle in the model has length, width, and area. For the first diagram shown in A1L94 (at the right), list the length, width, and area of each of the five rectangles below

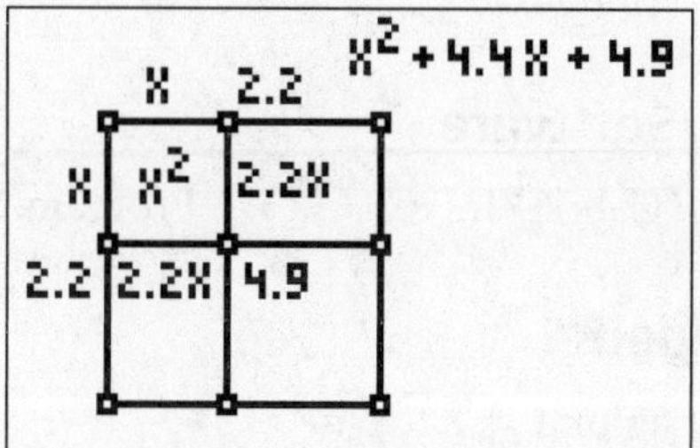

Rectangle	Length	Width	Area
top left			
top right			
bottom left			
bottom right			
large square			

2. The large figure above is a square. Given that the top-left rectangle in the figure is also a square, what must be true about the bottom-right rectangle? About the other two rectangles? Justify each answer.

3. You can change the dimensions of the large square by dragging the bottom-left vertex. Change the dimensions and complete the table below for your new model.

	Length	Width	Area
top left square			
each rectangle			
other inside square			
large square			

4. Predict the values you would get for each of the following when the two inside squares have lengths x and 2.5. Then test your predictions.

Area of each rectangle   Area of large square

Extension

5. Use pencil and paper. Generalize the model by drawing a diagram and making a table for which the large square has length $x + a$.

Square of a Binomial

Teacher Notes

Activity Objective

Students use Cabri® Jr. to develop understanding of how to square a binomial.

Time

- 15–20 minutes

Materials/Software

- App: Cabri® Jr.
- Program: A1L94
- Activity worksheet

Skills Needed

- drag a point

Classroom Management

- Students can work individually or in pairs depending on the number of calculators available.
- Use TI Connect™ software, TI-GRAPH LINK™ software, the TI-Navigator™ system, or unit-to-unit links to transfer A1L94 to each calculator.

Notes

- Remind students that Cabri® Jr. rounds values to the nearest tenth.

Answers

1.

Rectangle	Length	Width	Area
top left	x	x	x^2
top right	2.2	x	$2.2x$
bottom left	x	2.2	$2.2x$
bottom right	2.2	2.2	4.9
large rectangle	$x + 2.2$	$x + 2.2$	$x^2 + 4.4x + 4.9$

2. It is a square. The other two rectangles are congruent. Check students' work.

3. Check students' work.

4. $2.5x$; $x^2 + 5x + 6.25$

5.

	Length	Width	Area
top left square	x	x	x^2
each rectangle	a	x	ax
other inside square	a	a	a^2
large square	$x + a$	$x + a$	$x^2 + 2ax + a^2$

Name ______________________ Class ______________ Date ______________

Quadratic Graphs I

Activity 17

FILES NEEDED: Transformation Graphing App
Program: A1L101

A1L101 graphs the quadratic function $y = ax^2 + c$ as Y1 = AX^2 + C.

In this activity, you will explore the graph of $y = ax^2 + c$.

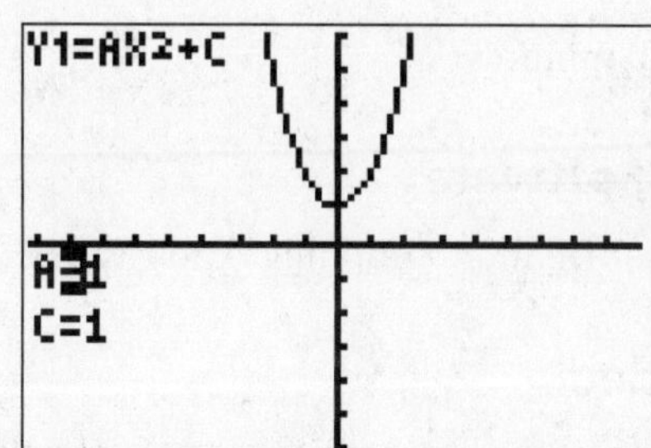

1. Run A1L101. Set Step = .2 and increment the value of A in steps of 1. When A > 0, what happens to the graph as A increases?

2. Decrease the value of A. What happens to the graph as A decreases? What happens to the graph as A gets close to zero but remains greater than zero?

3. Set A = −0.5. Describe the graph.

4. Try several different negative values for A. What happens to the graph as A gets close to zero?

5. For what values of A does the parabola open up? Open down?

6. What happens to the parabola when A = 0? Explain why this makes sense, given your answers to Questions 2 and 4.

Set A = 1, Step = 1, and select C.

7. Increase the value of C. What happens to the graph as C increases?

8. Predict what will happen to the graph as C decreases. Test your prediction.

Extension

The *vertex* is the highest or lowest point on the parabola.

9. What are the coordinates of the vertex of the parabola with A = 2 and C = −3? Test your guess.

10. Choose values of A and C such that the vertex of the parabola is the highest point and is located at (0, 5). Test your choices.

11. Find quadratic functions whose graphs are like those in this activity, except that their vertexes are at different places along the *x*-axis.

Quadratic Graphs I

Teacher Notes

Activity Objective

Students use the Transformation Graphing App to explore the graph of the parabola for the $y = ax^2 + c$ form.

Time

- 15–20 minutes

Materials/Software

- Transformation Graphing App
- Program: A1L101
- Activity worksheet

Skills Needed

- change parameter values
- change Step values

Classroom Management

- Students can work individually or in pairs depending on the number of calculators available.
- Encourage students to discuss their observations with other students.

Notes

- Students can enter values for A and B directly by typing the number and pressing ENTER.
- Students should uninstall the Transformation Graphing App at the end of this activity and then run DEFAULT to reset their calculators.
- Press WINDOW and change Step values in the SETTINGS menu.

Answers

1. The parabola gets narrower.

2. The parabola gets wider.

3. The parabola is wide and opens downward.

4. The parabola flattens upward.

5. For A > 0, the parabola opens up. For A < 0, the parabola opens down.

6. It becomes the horizontal line Y = 0. Since A = 0 is neither positive nor negative, the parabola does not open up *or* down.

7. The parabola moves up.

8. The parabola will move down.

9. (0, −3)

10. Answers may vary. Sample: A = −2, C = 5

11. Check students' work.

Name ______________________ Class ______________ Date ______________

Quadratic Graphs II

Activity 18

FILES NEEDED: Transformation Graphing App Program: A1L102A

A1L102A graphs the quadratic function $y = ax^2 + bx$ as $Y1 = AX^2 + BX$.

In this activity, you will explore the graph of $y = ax^2 + bx$. The program A1L102A includes a grid that shows a dot at each integer in the x- and y-directions.

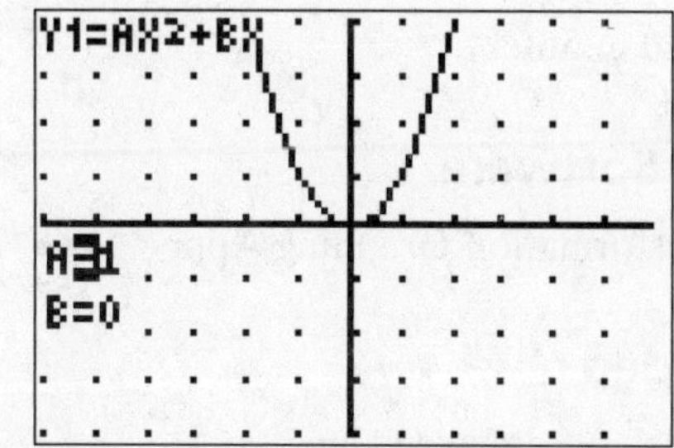

1. Run A1L102A. Select B and increment it in steps of 1. What happens to the vertex of the parabola as B increases?

2. Set B = −4. Complete the table below.

A	1	1	1	1	1
B	−4	−2	0	2	4
x-coordinate of vertex					

3. Use what you observed in Questions 1 and 2. How far does the vertex move horizontally each time you increase B by 1?

4. Set A = 2 and B = −4. Complete the table below.

A	2	2	2	2	2
B	−4	−2	0	2	4
x-coordinate of vertex					

5. Use what you observed in Question 4. How far does the vertex move horizontally each time you increase B by 1 when A = 2?

6. Complete this conjecture.

For $y = ax^2 + bx$, the x-coordinate of the vertex of the graph is $\frac{b}{■}$.

Extension

7. Use your ■ value from Question 6. Discuss the graph for ■ = 0.

8. Set A = 1 and B = −4. As you increment B, follow the path of the vertex. What kind of curve does it appear to trace? Give a convincing argument to support your answer.

Quadratic Graphs II

Teacher Notes

Activity Objective

Students use the Transformation Graphing App to explore the graph of the function $y = ax^2 + bx$ and the coordinates of the vertex of the parabola.

Time

- 15–20 minutes

Materials/Software

- Transformation Graphing App
- Program: A1L102A
- Activity worksheet

Skills Needed

- change parameter values

Classroom Management

- Students can work individually or in pairs depending on the number of calculators available.
- Use TI Connect™ software, TI-GRAPH LINK™ software, the TI-Navigator™ system, or unit-to-unit links to transfer A1L102A to each calculator.

Notes

- Students can enter values for A and B directly by typing the number and pressing ENTER.
- Students should uninstall the Transformation Graphing App at the end of this activity and then run DEFAULT.

Answers

1. It moves to the left and down. **2.** 2; 1; 0; −1, −2 **3.** $-\frac{1}{2}$ unit

4. 1; $\frac{1}{2}$; 0; $-\frac{1}{2}$; 1 **5.** $-\frac{1}{4}$ unit **6.** $-2a$

7. If $-2a = 0$, then $a = 0$, and $y = bx$, which is the equation of a line.

8. Answers may vary. Sample: Since A is constant, the shape of the parabola does not change. C = 0 so the parabola always passes through (0, 0). The effect is to view the parabola as a solid object that is sliding through an opening at the origin. For $Y1 = X^2 + BX$, the vertex is at $X = \frac{-B}{2}$, so the vertex follows the path of $Y1 = \left(\frac{-B}{2}\right)^2 + B\left(\frac{-B}{2}\right) = -\frac{1}{4}B^2$, a parabola that opens down.

Name ______________________ Class ______________ Date ______________

Quadratic Function Match I

Activity 19

FILES NEEDED: Transformation Graphing App
Program: A1L102B

A1L102B graphs the quadratic function $y = ax^2 + bx$ as $Y1 = AX^2 + BX$ with $A = 1$ and $B = 0$.

In this activity you will see three plots, one at a time. You are to change the A and B values for $Y1 = AX^2 + BX$ until you find a function $y = ax^2 + bx$ that is a perfect match for the given plot.

1. Run A1L102B. Find a quadratic function whose graph matches the given plot, shown at the right.

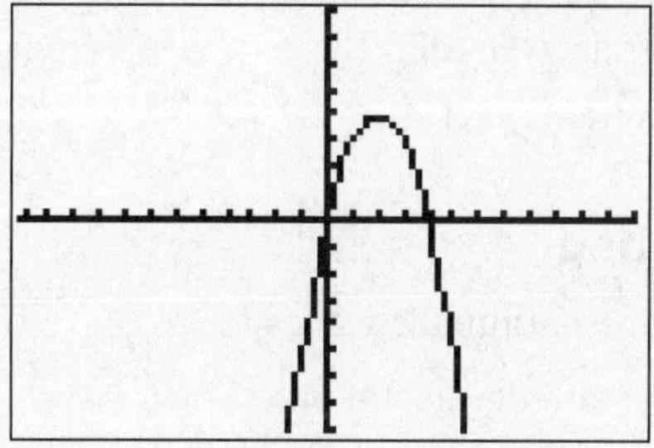

2. Switch from Plot1 to Plot2. Press GRAPH to see A1L102B for the second plot. Find a matching quadratic function.

Plot1 Plot2 Plot3
\Y1=AX²+BX
\Y2=
\Y3=
\Y4=
\Y5=
\Y6=
\Y7=

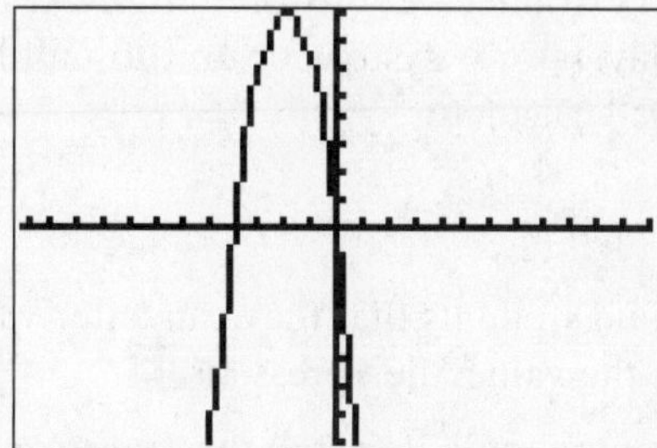

3. Switch from Plot2 to Plot3. Press GRAPH to see A1L102B for the third plot. Find a matching quadratic function.

4. By looking at the graphs, how can you tell that none of the functions you seek would have the form $y = ax^2 + c$?

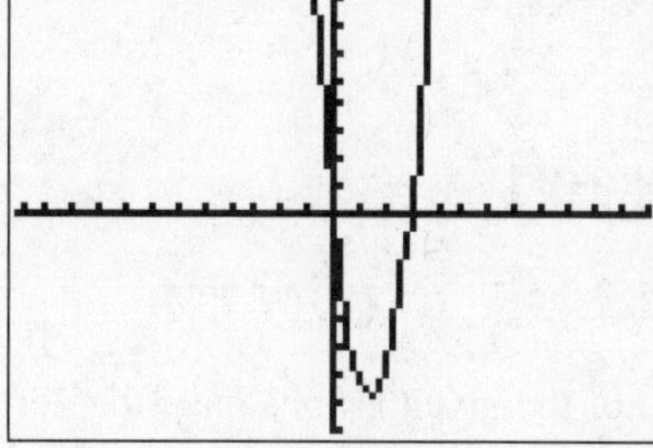

Quadratic Function Match I

Teacher Notes

Activity Objective

Students use the Transformation Graphing App to learn how changing the values of a and b affects the graph of $y = ax^2 + bx$.

Time

- 15–20 minutes

Materials/Software

- Transformation Graphing App
- Program A1L102B
- Activity worksheet

Skills Needed

- change parameter values
- select and deselect a plot

Classroom Management

- Students can work individually or in pairs depending on the number of calculators available.
- Use TI Connect™ software, TI-GRAPH LINK™ software, the TI-Navigator™ system, or unit-to-unit links to transfer A1L102B to each calculator.

Notes

- Remind students that they can enter values directly for A= or B=. Type the value; then press ENTER.
- Students should uninstall the Transformation Graphing App at the end of this activity and then run DEFAULT.

Answers

1. $y = -x^2 + 4x$
2. $y = -2x^2 - 8x$
3. $y = 3x^2 - 9x$
4. None of the given graphs have the vertex on the y-axis. To have $y = ax^2 + c$ form, the vertex would be on the y-axis.

Name ______________________ Class ______________ Date ______________

Graphs, Solutions, and Factors

Activity 20

> **FILES NEEDED:** Transformation Graphing App
> Program: A1L105

In A1L105, the quadratic function $y = (x - a)(x - b)$ is in *factored form* and is graphed as Y1 = (X – A)(X – B).

In this activity, you will explore relationships among

- the x-intercepts of a quadratic function,
- the solutions of the related quadratic equation, and
- the factors of the related quadratic expression.

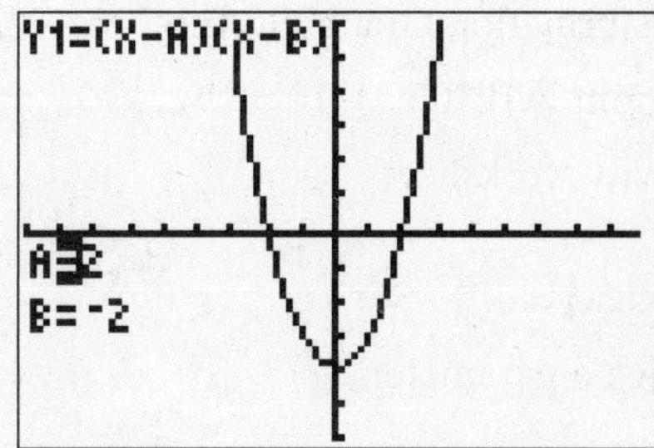

1. Run A1L105. For the graph that appears, what are the initial values given for A and B? What are the x-intercepts of the graph?

2. Change the value of A. What happens to the values of the x-intercepts of the graph when you change A? Change the value of A again to check what you have observed.

3. Set A = 2. Select B and change its value. What happens to the values of the x-intercepts when you change B? Change the value of B again to check what you have observed.

4. Complete this conjecture.

The x-intercepts of the graph of $y = (x - a)(x - b)$ are __?__.

5. Use your answer to Question 4 to predict the x-intercepts of the graph of the quadratic function $y = (x - 1)(x - 5)$. Set A = 1 and B = 5 to check your prediction.

6. The quadratic function $y = (x - a)(x - b)$ has $(x - a)(x - b) = 0$ as its *related quadratic equation*. Explain why the x-intercepts of the graph of the quadratic function are the solutions of the related equation.

7. The quadratic function $y = x^2 - 5x + 6$ is in standard form. Write it in factored form. Predict the x-intercepts of its graph and the solutions of the equation $x^2 - 5x + 6 = 0$. Use A1L105 to check your predictions.

Extension

8. Describe what happens to the graph of $y = (x - a)(x - b)$ as you bring the values of A and B closer together.

9. When A = B, what appears to be true about the graph? What does this mean in terms of solutions of the related quadratic equation?

Graphs, Solutions, and Factors

Teacher Notes

Activity Objective

Students use the Transformation Graphing App to explore the relationships among the x-intercepts of a quadratic function, the solutions of the related quadratic equation, and the factors of the related quadratic expression.

Time

- 15–20 minutes

Materials/Software

- Transformation Graphing App
- Program: A1L105
- Activity worksheet

Skills Needed

- change a parameter

Classroom Management

- Students can work individually or in pairs depending on the number of calculators available.
- Use TI Connect™ software, TI-GRAPH LINK™ software, the TI-Navigator™ system, or unit-to-unit links to transfer A1L105 to each calculator.

Notes

- Students should uninstall the Transformation Graphing App at the end of this activity and then run DEFAULT to reset their calculators.

Answers

1. 2, −2; 2, −2
2. One x-intercept moves with the value of A.
3. The other x-intercept moves with the value of B.
4. a, b
5. 1, 5
6. If $x = a$ then $(x - a) = 0$ and $(x - a)(x - b) = 0$. Likewise, if $x = b$ then $(x - b) = 0$ and $(x - a)(x - b) = 0$.
7. $y = (x - 3)(x - 2)$; 3, 2
8. The x-intercepts move closer together. The vertex gets closer to the x-axis.
9. The vertex sits on the x-axis. There is only one solution to the related quadratic equation.

Name ______________________ Class ______________ Date ______________

The Discriminant

Activity 21

FILES NEEDED: Transformation Graphing App
Program: A1L108

In A1L108, the quadratic function $y = ax^2 + bx + c$ is in standard form. It is graphed as $Y1 = AX^2 + BX + C$. The related quadratic equation is $ax^2 + bx + c = 0$.

In this activity, you will explore the relationship between the value of $b^2 - 4ac$ and the number of solutions of the related quadratic equation.

1. Recall: How do the number of x-intercepts of the graph of a quadratic function relate to the number of solutions of the corresponding quadratic equation? (*Hint:* What is the value of the function at an x-intercept?)

2. Run A1L108. Find four sets of values for A, B, and C so the parabola intersects the x-axis exactly once. Record A, B, C, and the values of $B^2 - 4AC$ in the table below.

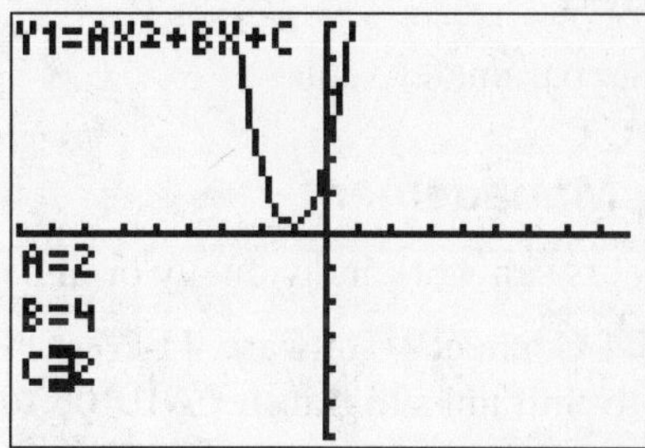

A				
B				
C				
$B^2 - 4AC$				

3. Study the data in the table. Complete this conjecture about the relationship between a quadratic equation with one solution and the value of $b^2 - 4ac$.

The graph of $y = ax^2 + bx + c$ has one x-intercept and the related equation $ax^2 + bx + c = 0$ has one solution if $b^2 - 4ac$ = __?__.

4. Find four sets of A, B, C values so the parabola does *not* intersect the x-axis. Record A, B, C, and the values of $B^2 - 4AC$ in the table below.

A				
B				
C				
$B^2 - 4AC$				

5. Study the data in the table. Make a conjecture about the relationship between a quadratic equation with no solutions and the value of $b^2 - 4ac$.

6. Study your conjectures from Questions 3 and 5. Make a conjecture about the relationship between a quadratic equation with two solutions and the value of $b^2 - 4ac$. Find values for a, b, and c that give a desired value of $b^2 - 4ac$. Then test your conjecture in A1L108.

The Discriminant

Activity Objective

Students use the Transformation Graphing App to discover the relationship between the number of solutions of a quadratic equation and the value of the discriminant.

Time

- 15–20 minutes

Materials/Software

- Transformation Graphing App
- Program: A1L108
- Activity worksheet

Skills Needed

- change parameter values

Classroom Management

- Students can work individually or in pairs depending on the number of calculators available.
- Use TI Connect™ software, TI-GRAPH LINK™ software, the TI-Navigator™ system, or unit-to-unit links to transfer A1L108 to each calculator.

Notes

- Remind students that seeing many cases of a situation does *not* prove that the situation is always true!
- Students should uninstall the Transformation Graphing App at the end of this activity and then run DEFAULT to reset their calculators.

Answers

1. They are the same.

2. Check students' work.

3. 0

4. Check students' work.

5. If a quadratic equation has no solutions, then $b^2 - 4ac < 0$.

6. If a quadratic equation has two solutions, then $b^2 - 4ac > 0$.

Name ______________________ Class ______________ Date ______________

Choose a Model I

Activity 22

FILES NEEDED: Transformation Graphing App Program: A1L109

In A1L109, the scatter plot suggests a relationship between x and y values.

In this activity, you will explore ways to use the Transformation Graphing App to help you find a reasonable model for the relationship.

1. A good first step is to recall the models you have learned:
- linear
- quadratic
- exponential

Then you think, "Which of a line, a parabola, or an exponential curve could fit the scatter plot very well?" Which two of these three types of functions would give a reasonably good answer? Explain.

The general forms of these models are

$y = mx + b \qquad y = ax^2 + bx + c \qquad y = ab^x$

On your graphing calculator, these could appear as

Y1 = AX + B Y2 = AX² + BX + C Y3 = AB^X

The numbers a, b, and c are called *parameters*. Choose two of Y1, Y2, and Y3 (match your choices from Question 1) to use for the following questions.

2. Run A1L109. Select one of your functions by highlighting the equal sign. Press GRAPH, then change its parameter values until graph points are close to given data points. Write the function for your graph.

3. As a check, find the Y-value for X = 5 and the Y-value of the data point at $x = 5$. How do they compare?

4. Select your second function by highlighting its equal sign. Change its parameter values until the points on the graph are close to the given data points. Write the function for this second graph.

5. Repeat the check of Question 3 for this function.

6. Which of the functions found in Questions 2 and 4 seem to work the best for the given data, or can you tell? You may find it helpful to highlight both of the equal signs and draw both graphs in the same window. Explain your choice.

Extension

7. Adjust the window settings to look at data points with x-values between 10 and 20. What can you conclude?

Choose a Model I

Teacher Notes

Activity Objective

Students use the Transformation Graphing App to compare quadratic and linear models.

Time

- 20–30 minutes

Materials/Software

- Transformation Graphing App
- Program: A1L109
- Activity worksheet

Skills Needed

- change parameter values
- select a function

Classroom Management

- Students can work individually or in pairs depending on the number of calculators available.
- Use TI Connect™ software, TI-GRAPH LINK™ software, the TI-Navigator™ system, or unit-to-unit links to transfer A1L109 to each calculator.

Notes

- Remind students that their models are likely to be different.
- Students may need to enter decimal values to hundredths to model the given data.
- Students can use value in the CALC menu to help answer Question 2.
- Students should uninstall the Transformation Graphing App at the end of this activity and then run DEFAULT to reset their calculators.

Answers

Answers may vary. Samples are given:

1. Quadratic or exponential model; their graphs are curved as is the scatter plot.

2. $y = 0.08x^2 - 1.2x + 4.9$ **3.** 0.9; 1.2 **4.** $y = 4.1(0.8)^x$ **5.** $\approx 1.3; 1.2$

6. Check students' work.

7. The quadratic function does not model the data well because its graph rises for $x > 10$ while the data y-values continue to decrease.

Name ______________________ Class ______________ Date ______________

Square Root Graphs

Activity 23

FILES NEEDED: Transformation Graphing App
Program: A1L116A

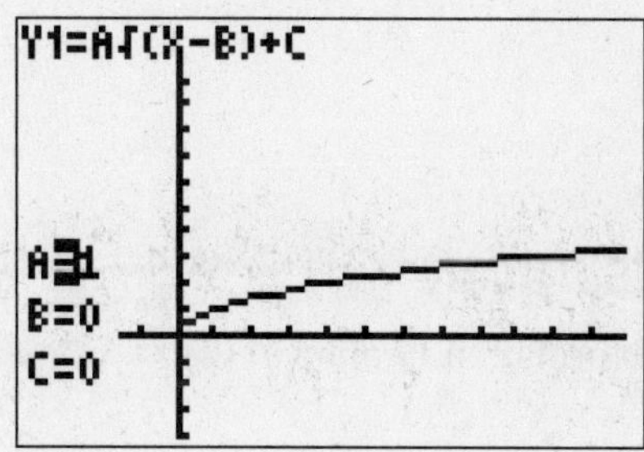

A1L116A graphs the square root function $y = a\sqrt{x - b} + c$ as Y1 = A$\sqrt{\ }$ (X – B) + C.

In this activity, you will explore the graph of the square root function.

1. Write this function with A = 1, B = 0, and C = 0. This function is sometimes called the square root function. What is its domain?

2. Run A1L116A. Increment the value of B in steps of 1 or −1. What happens to the graph as B increases? As B decreases?

3. Set B = 4. What is the domain of the function for B = 4?

4. Set B = −1. What is the domain of the function for B = −1?

5. Set B = 1 and then select C. Change the value of C. How does the value of C affect the graph of the function?

6. Set C = 3. What is the range of the function?

7. Predict the domain and range of $y = \sqrt{x - 2} + 4$. Test your predictions by setting B and C to the correct values in A1L116A.

8. Write a function that has domain $x \geq -2$ and range $y \geq 5$. Graph your function to check the domain and range.

9. Set A, B, and C to match the screen at the right. Increment the value of A in steps of 1. What effect does this have on the graph?

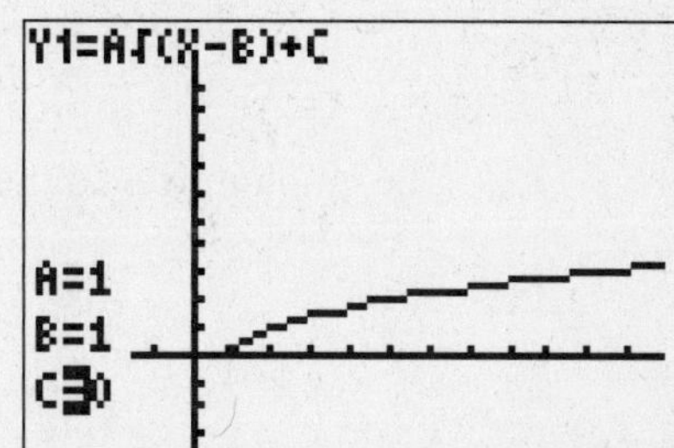

10. What happens to the graph when A < 0? (*Hint:* Try A = −0.5, A = −1, and A = −2.)

11. What happens if A = 0? Explain.

Extension

12. Each graph shown above is half of a parabola. For each screen, find a quadratic function whose graph, cut in half, has the same shape. (*Hint:* It will open upwards or downwards.) Justify your answer.

13. Describe how the effects of changing a, b, and c in $y = a\sqrt{x - b} + c$ and in $y = a|x - b| + c$ are similar.

Square Root Graphs

Teacher Notes

Activity Objective

Students use the Transformation Graphing App to explore how values of a, b, and c affect the graph of $y = a\sqrt{x - b} + c$.

Time

- 15–20 minutes

Materials/Software

- Transformation Graphing App.
- Program: A1L116A
- Activity worksheet

Skills Needed

- change parameter values

Classroom Management

- Students can work individually or in pairs depending on the number of calculators available.
- Use TI Connect™ software, TI-GRAPH LINK™ software, the TI-Navigator™ system, or unit-to-unit links to transfer A1L116A to each calculator.

Notes

- Use the term "translation" when appropriate.
- Remind students that the form has a "$- b$," so when they are asked to discuss $y = a\sqrt{x - 2} + 4$ in Question 7, $b = 2$.
- Students should uninstall the Transformation Graphing App at the end of this activity and then run DEFAULT to reset their calculators.

Answers

1. $x \geq 0$ **2.** The graph moves to the right. The graph moves to the left.

3. $x \geq 4$ **4.** $x \geq -1$ **5.** It translates the graph up or down.

6. $y \geq 3$ **7.** $x \geq 2, y \geq 4$ **8.** $y = \sqrt{x + 2} + 5$

9. As A increases, the graph rises more steeply.

10. The graph "flips." **11.** The graph becomes the horizontal line $y = 0$.

12. Answers may vary. Samples: first screen, $y = x^2$; second screen, $y = x^2 + 1$

13. a: stretches vertically; b: shifts each left or right; c: shifts each up or down.

Name ____________________ Class ____________ Date ____________

Square Root Function Match

Activity 24

FILES NEEDED: Transformation Graphing App
Program: A1L116B

A1L116B graphs the square root function $y = a\sqrt{x - b} + c$ as Y1 = A√ (X − B) + C with A = 1, B = 0, and C = 0.

In this activity you will see three plots, one at a time. You are to change the A , B, and C values for Y1= A√ (X − B) + C until you find a function $y = a\sqrt{x - b} + c$ that is a perfect match for the given plot.

1. Run A1L116B. Find a square root function whose graph matches the given plot, shown at the right.

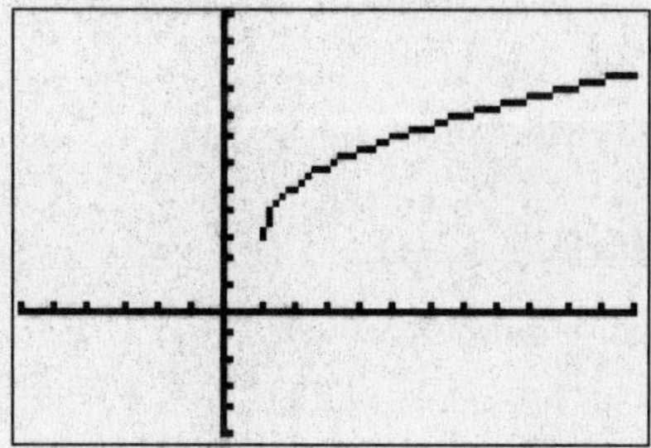

2. Switch from Plot1 to Plot2. Press GRAPH to see A1L116B for the second plot. Find a matching square root function.

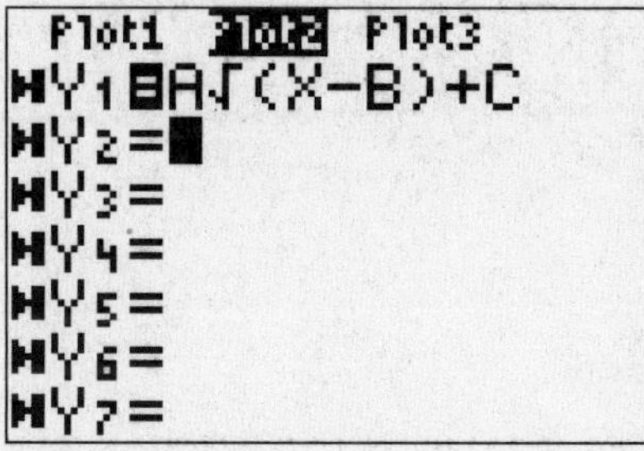

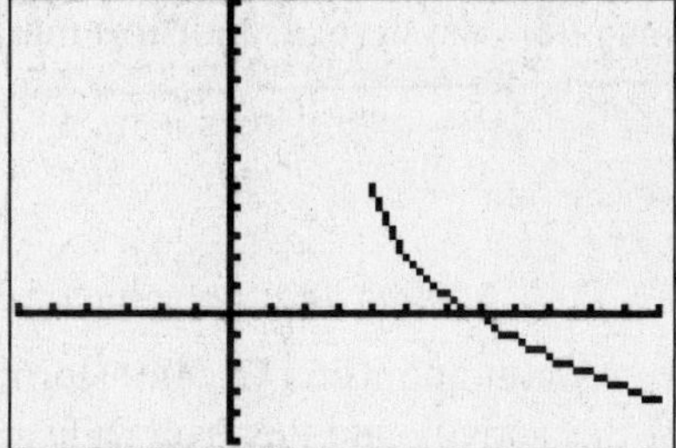

3. Switch from Plot2 to Plot3. Press GRAPH to see A1L116B for the third plot. Find a matching square root function.

4. By looking at the graphs, how can you tell whether to use positive values for A or negative values for A?

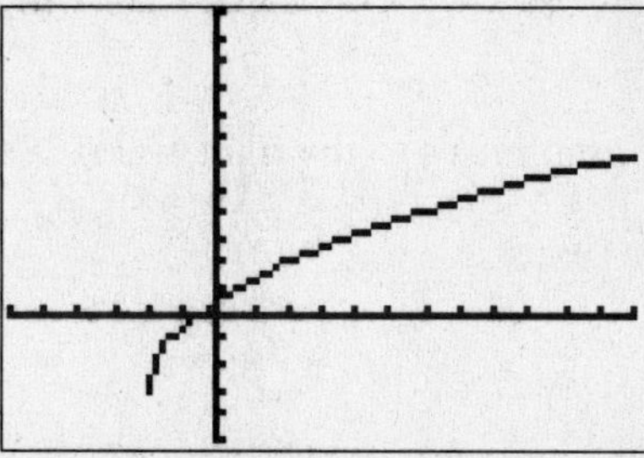

Square Root Function Match

Teacher Notes

Activity Objective

Students use the Transformation Graphing App to learn how changing the values of a, b, and c affects the graph of $y = a\sqrt{x - b} + c$.

Time

- 15–20 minutes

Materials/Software

- Transformation Graphing App
- Program: A1L116B
- Activity worksheet

Skills Needed

- change parameter values
- select and deselect plots

Classroom Management

- Students can work individually or in pairs depending on the number of calculators available.
- Use TI Connect™ software, TI-GRAPH LINK™ software, the TI-Navigator™ system, or unit-to-unit links to transfer A1L116B to each calculator.

Notes

- Suggest that students find the value of B, then C, then A.
- Students should uninstall the Transformation Graphing App at the end of this activity and then run DEFAULT.

Answers

1. $y = 2\sqrt{x - 1} + 3$ **2.** $y = -3\sqrt{x - 4} + 5$ **3.** $y = 2.5\sqrt{x + 2} - 3$

4. If the graph rises to the right, use $A > 0$. If the graph falls to the right, use $A < 0$.

Name ______________________ Class ______________ Date ______________

Rational Graphs

Activity 25

FILES NEEDED: Transformation Graphing App
Program: A1L122A

A1L122A graphs the rational function $y = \frac{a}{x - b} + c$ as Y1 = (A/(X − B)) + C with A = 4, B = 0, and C = 0.

In this activity you will explore the graph of the rational function.

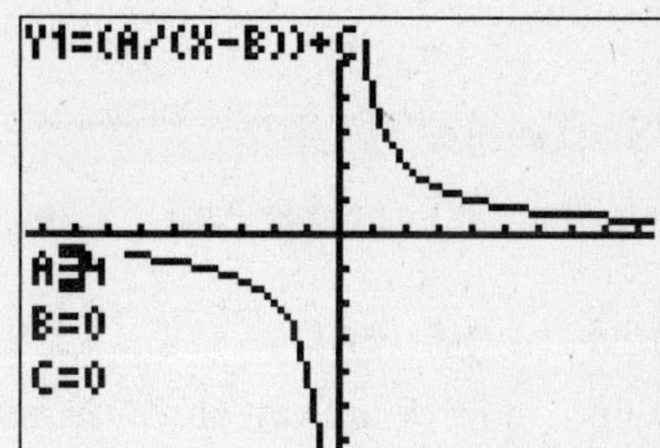

1. Run A1L122A. Change the value of A. What is the effect on the graph as A increases? As A decreases?

2. Predict the effects that you think the values B and C have on the graph of $y = \frac{a}{x - b} + c$. [*Hint:* Review the effects of B and C in "Absolute Value Graphs" (p. 13) and "Square Root Graphs" (p. 45).]

As $|x|$ gets large, the graph of the rational function $y = \frac{a}{x}$ $(b = 0, c = 0)$ approaches a horizontal line (the x-axis in this case). As $|x|$ gets small, the graph approaches a vertical line (here, the y-axis). Any rational function has two such *asymptotes*. On the screen, you generally will not see them. On paper, you should draw the asymptotes to help you graph the function.

In the Questions that follow, observe (in your mind's eye) the asymptotes as you change values for B and C.

3. Increase the value of B to 2. What happens to the graph? To the asymptotes?

4. How does changing the value of B affect the graph? The asymptotes?

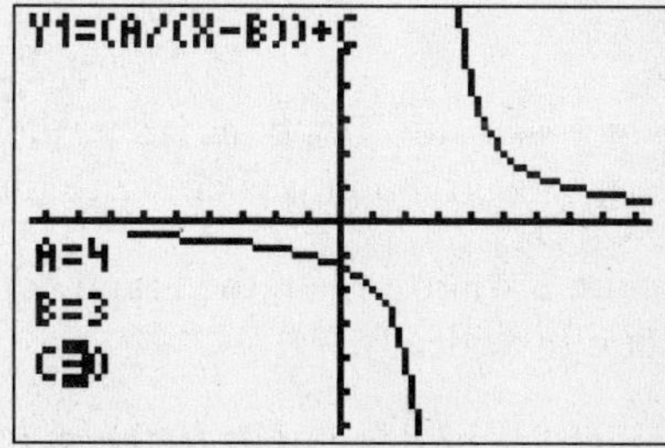

5. Predict the value for B that will put the vertical asymptote at –3? Test your prediction.

6. Set B = 3. Increase the value of C. Describe the effect of C on the graph?

7. Where is the horizontal asymptote for C = 4?

8. What equation will have a vertical asymptote at $x = -3$ and a horizontal asymptote at $y = 2$? Test your equation.

Extension

9. What type of variation does a rational function model when $b = 0$, and $c = 0$?

10. What happens to a rational function graph when you replace A by its opposite?

Rational Graphs

Teacher Notes

Activity Objective

Students use the Transformation Graphing App to explore how changing the values of a, b, and c affects the graph of $y = \frac{a}{x - b} + c$.

Time

- 25–30 minutes

Materials/Software

- Transformation Graphing App
- Program: A1L122A
- Activity worksheet

Classroom Management

- Students can work individually or in pairs depending on the number of calculators available.
- Use TI Connect™ software, TI-GRAPH LINK™ software, the TI-Navigator™ system, or unit-to-unit links to transfer A1L122A to each calculator.

Notes

- Remind students that when they graph by hand, they should first locate and graph the asymptotes.
- Discuss why there are two asymptotes.
- Compare the translations of this function with the translations of the functions previously studied.
- Remind students that in Question 8 the denominator becomes $x - (-3)$, or $x + 3$.

Answers

1. As A increases, the graph moves farther from the origin. As A decreases the graph moves closer to the origin.

2. Changing B will translate the graph horizontally. Changing C will translate it vertically.

3. The graph translates 2 units to the right. The vertical asymptote becomes $x = 2$. The horizontal asymptote remains $y = 0$.

4. Changing B translates the graph and the vertical asymptote horizontally.

5. B = –3

6. Changing C translates the graph vertically.

7. $y = 4$

8. $y = \frac{a}{x + 3} + 2$

9. inverse variation

10. The graph reflects across the horizontal asymptote.

Name ______________________ Class ______________ Date ______________

Rational Function Match

Activity 26

FILES NEEDED: Transformation Graphing App
Programs: A1L122B, A1L122C, A1L122D

Each program above graphs the rational function $y = \frac{a}{x - b} + c$ as Y1 = A/(X – B) + C with A = 2, B = 0, and C = 0.

The startup screen shows two asymptotes and plots 6 points for the graph of a rational function. You are to change the values of A, B and C until you find a function $y = \frac{a}{x - b} + c$ that is a perfect match for the given plot.

1. Run A1L122B. Find a rational function whose graph has the given asymptotes and contains the six given points.

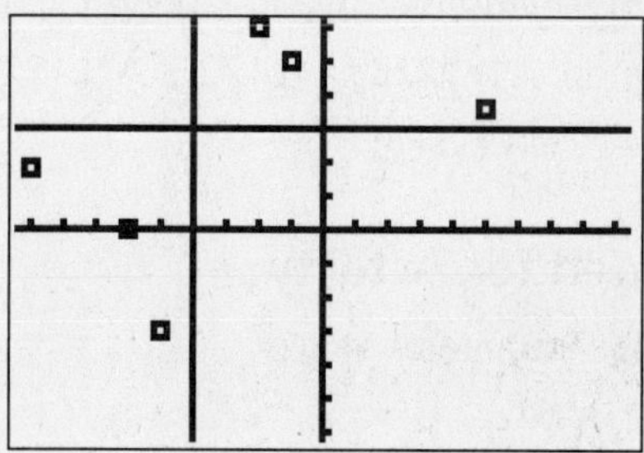

2. Run A1L122C. Find a rational function whose graph has the given asymptotes and contains the six given points.

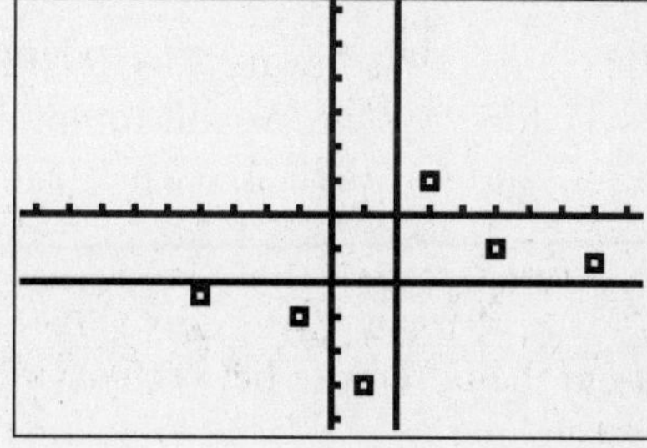

3. Run A1L122D. Find a rational function whose graph has the given asymptotes and contains the six given points.

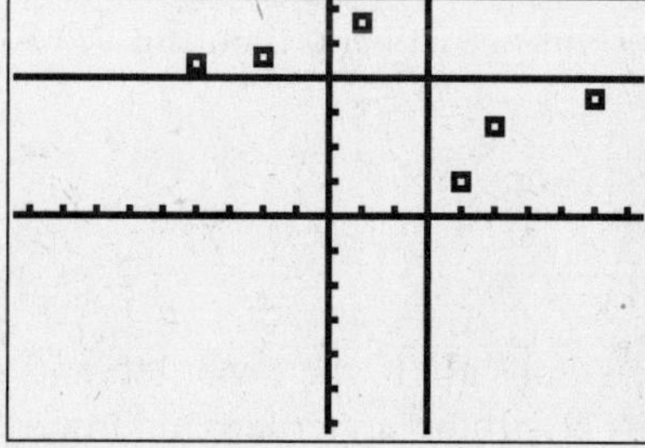

4. By looking at the graphs, how can you tell whether to use positive values for A or negative values for A?

Rational Function Match

Teacher Notes

Activity Objective

Students use the Transformation Graphing App to study how changing parameter values affects the graph of $y = \frac{a}{x - b} + c$.

Time

- 15–20 minutes

Materials/Software

- Transformation Graphing App
- Program: A1L122B, A1L122C, A1L122D
- Activity worksheet

Skills Needed for Activity

- change parameter values

Classroom Management

- Students can work individually or in pairs depending on the number of calculators available.
- Use TI Connect™ software, TI-GRAPH LINK™ software, the TI-Navigator™ system, or unit-to-unit links to transfer A1L122B, A1L122C, and A1L122D to each calculator.

Notes

- Point out the difference between asymptotes (no scale marks) and axes (scale marks).
- Encourage students to determine the value of B, then C, then A.
- Students should uninstall the Transformation Graphing App at the end of this activity and then run DEFAULT.

Answers

1. $y = \frac{6}{x + 4} + 3$ **2.** $y = \frac{3}{x - 2} - 2$ **3.** $y = \frac{-3}{x - 3} + 4$

4. If the graphs are to the lower left and upper right of the asymptotes, use $A > 0$. If the graphs are to the upper left and lower right of the asymptotes, use $A < 0$.

Name ______________________ Class ______________ Date ______________

Perpendicular Bisectors

Activity 27

FILES NEEDED: Cabri® Jr.
AppVar: GL15A

Given: In GL15A, point T is on line 1, the perpendicular bisector of $\overline{AB}$.

Explore: properties of the perpendicular bisector

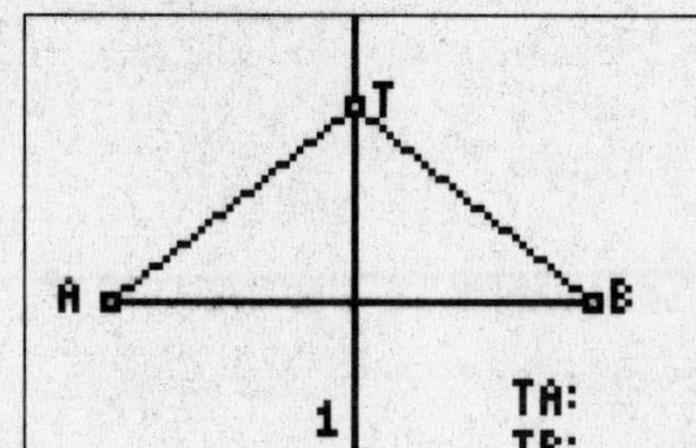

1. Drag point T along line 1. Find lengths TA and TB for four different locations of T. Collect your data in the table below.

Length TA				
Length TB				

2. Drag point A or B to change the length of $\overline{AB}$. Find lengths TA and TB for four lengths of $\overline{AB}$. Collect your data in the table below.

Length TA				
Length TB				

3. Study the data in each table. Complete this conjecture about how lengths TA and TB are related.

If point T is on the perpendicular bisector of segment $\overline{AB}$, then $TA = \underline{\ ?\ }$.

4. Generalize your conjecture from Question 3. Use just one word for the blank below.

If a point is on the perpendicular bisector of a segment, then the point is __?__ from the endpoints of the segment.

Extension

5. Suppose you have a point Q in the diagram so that $QA = QB$. Make a conjecture about the location of point Q. Then test your conjecture. Draw point Q. Install measures QA and QB. Then drag Q to make $QA = QB$.

6. Suppose you have two segments, $\overline{AB}$ and $\overline{CD}$, not in the same line. Make a conjecture about the location of a point Q for which $QA = QB$ and $QC = QD$. Test your conjecture.

Perpendicular Bisectors

Teacher Notes

Activity Objective

Students use Cabri® Jr. to explore the properties of perpendicular bisectors.

Time

- 10–15 minutes

Materials/Software

- App: Cabri® Jr.
- AppVar: GL15A
- Activity worksheet

Skills Needed

- drag an object
- install a measure

Classroom Management

- Students can work individually or in pairs depending on the number of calculators available.
- Use TI Connect™ software, TI-GRAPH LINK™ software, the TI-Navigator™ system, or unit-to-unit links to transfer GL15A to each calculator.

Notes

- Reminder: As noted on p. vi, each Activity page assumes that you activate the appropriate App at the start of the activity.
- In F1, select Open and then press ENTER to see the AppVar list.
- Dragging point A or B will change the orientation of the diagram but will not affect the perpendicular bisector relationship.
- If students change the length of $\overline{AB}$ by dragging either Point A or Point B, the length TA will still equal the length TB.

Answers

1–2. Check students' work. **3.** TB **4.** equidistant

5. Point Q must be on line l.

6. Q is on the perpendicular bisector of both $\overline{AB}$ and $\overline{CD}$.

Name ______________________ Class ______________ Date ______________

Angle Bisectors

Activity 28

FILES NEEDED: Cabri® Jr. AppVar: GL15B

Given: In GL15B, $\overline{PR}$ is the bisector of $\angle CPD$. Point T is on $\overline{PR}$. Points A and B are on the sides of the angle. $\overline{TA}$ and $\overline{TB}$ are perpendicular to the sides.

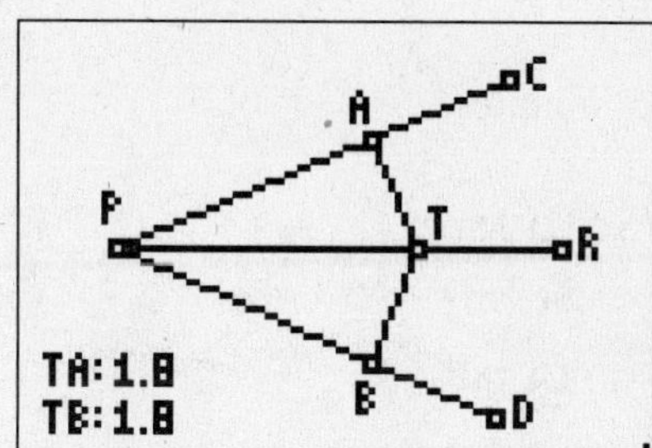

Explore: the properties of the angle bisector

1. Drag point T along $\overline{PR}$. Find lengths TA and TB for four different locations of T. Collect your data in the table below.

Length TA				
Length TB				

Open GL15B again. This returns you to the original figure.

2. Drag point C or D to change the size of $\angle CPD$. Find lengths TA and TB for four sizes of $\angle CPD$. Collect your data in the table below.

Length TA				
Length TB				

3. Study the data in each table. Complete this conjecture about how lengths TA and TB are related.

If point T is on the bisector of $\angle CPD$, and $\overline{TA}$ and $\overline{TB}$ are perpendicular to the sides of the angle at points A and B, respectively, then TA = __?__.

4. The length of the perpendicular segment from a point to a line is the *distance from the point to the line*. Use this to help you generalize your conjecture from Question 3. Use just one word for the blank below.

A point on the bisector of an angle is __?__ from the sides of the angle.

Extension

5. Suppose you add a point Z to the diagram with the property that the lengths of the perpendicular segments from Z to sides $\overline{PC}$ and $\overline{PD}$ are equal. Make a conjecture about the location of point Z. Test your conjecture.

Angle Bisectors

Teacher Notes

Activity Objective

Students use Cabri® Jr. to explore the properties of angle bisectors.

Time

- 10–15 minutes

Materials/Software

- App: Cabri® Jr.
- AppVar GL15B
- Activity worksheet

Skills Needed

- drag an object
- install a measure

Classroom Management

- Students can work individually or in pairs depending on the number of calculators available.
- Use TI Connect™ software, TI-GRAPH LINK™ software, the TI-Navigator™ system, or unit-to-unit links to transfer GL15B to each calculator.

Notes

- In F1, select Open and then press ENTER to see the AppVar list.
- Segments are used rather than rays because Cabri® Jr. does not support the construction of rays.
- Students can change $\angle CPD$ by dragging points C, P, or D.

Answers

1–2. Check students' work.

3. TB

4. equidistant

5. Point Z is on angle bisector $\overline{PR}$.

Name ______________________ Class ______________ Date ______________

Linear Pairs

Activity 29

FILES NEEDED: Cabri® Jr.
AppVar: GL24

Given: In GL24, Point O is on $\overleftrightarrow{AC}$. Point B is not on $\overleftrightarrow{AC}$. $\angle AOB$ and $\angle BOC$ are a linear pair of angles.

Explore: the angle measures of a linear pair

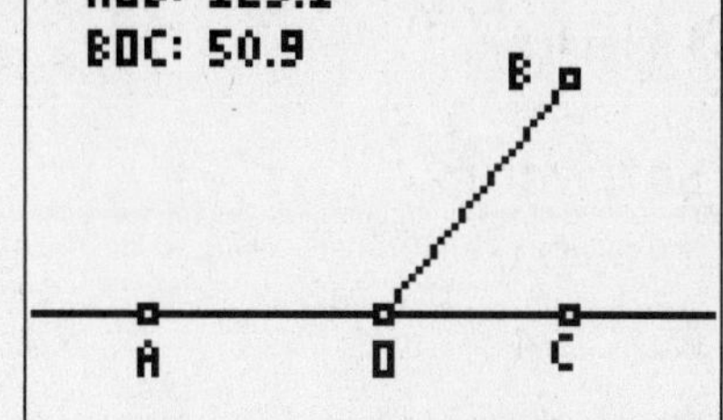

1. Drag point B to four different locations. Find $m\angle AOB$ and $m\angle BOC$ for each location. Collect your data in the table below.

2. For each location, find the sum of the measures of $\angle AOB$ and $\angle BOC$.

$m\angle AOB$				
$m\angle BOC$				
$m\angle AOB + m\angle BOC$				

3. Study the data in the table. Complete this conjecture about the sum of the measures of $\angle AOB$ and $\angle BOC$.

   If $\angle AOB$ and $\angle BOC$ are a linear pair, then __?__ .

4. Generalize your conjecture from Question 3.

   If two angles form a linear pair, then __?__ .

Extension

5. Drag point B so that $m\angle AOB = m\angle BOC$. What must be true about $\overline{OB}$ and $\overleftrightarrow{AC}$? Explain.

6. Drag B so that $m\angle BOC = 3\ m\angle BOC$. What are the measures of each angle? Verify your answer using algebra.

7. Write a problem similar to Question 6 that can be answered by dragging B and then verified algebraically. Give your question to a classmate to answer and verify.

Linear Pairs

Teacher Notes

Activity Objective

Students use Cabri® Jr. to explore properties of linear pairs.

Time

- 10–15 minutes

Materials/Software

- App: Cabri® Jr.
- AppVar: GL24
- Activity worksheet

Skills Needed

- drag an object

Classroom Management

- Students can work individually or in pairs depending on the number of calculators available.
- Use TI Connect™ software, TI-GRAPH LINK™ software, the TI-Navigator™ system, or unit-to-unit links to transfer GL24 to each calculator.

Notes

- In F1, select Open and then press ENTER to see the AppVar list.
- If the sums of the angle measures stay close to the same value, they should be considered equal.

Error Prevention

- Cabri® Jr. will not measure angles greater than 180°. Students should keep point B above $\overline{AC}$.

Answers

1–2. Check students' work.

3. $m\angle AOB + m\angle BOC = 180$

4. their sum is 180

5. They are perpendicular. If $x + y = 180$ and $x = y$, then $x = 90$.

6. $m\angle AOB = 135, m\angle BOC = 45$. If $x + y = 180$, and $x = 3y$, then $3y + y = 180, 4y = 180, y = 45$, and $x = 135$.

7. Check students' work.

Name ______________________ Class ______________ Date ______________

Vertical Angles

Activity 30

FILES NEEDED: Cabri® Jr.
AppVar: GL25

Given: In GL25, $\overleftrightarrow{AC}$ and $\overleftrightarrow{BD}$ intersect at point O.

Explore: the angle measures of vertical angles

1. Drag point A or point B to four different locations. Find $m\angle AOB$, $m\angle BOC$, $m\angle COD$, and $m\angle AOD$ for each location. Collect your data in the table below.

$m\angle AOB$				
$m\angle BOC$				
$m\angle COD$				
$m\angle AOD$				

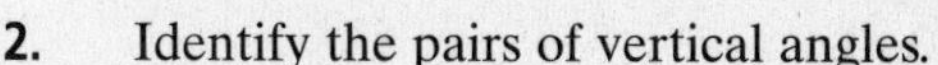

2. Identify the pairs of vertical angles.

3. Study the data in the table. Complete this conjecture about vertical angles.

If $\angle AOB$ and __?__ are vertical angles, then __?__ .

4. Generalize your conjecture from Question 3.

If two angles are vertical angles, then __?__ .

Extension

5. Drag point A or point B so that $m\angle AOB = m\angle BOC$. What must be true about $\overleftrightarrow{AC}$ and $\overleftrightarrow{BD}$?

Suppose you construct a third line as described. How many pairs of vertical angles can you find?

6. $\overleftrightarrow{EF}$ containing point O

7. $\overleftrightarrow{EF}$ containing point B

Animation Option

Instead of dragging point A or B in Question 1, animate A or B. Then stop the point at different locations to collect your data. (You can reset the animation by using Undo in the F1 menu.)

Vertical Angles

Teacher Notes

Activity Objective

Students use Cabri® Jr. to explore the properties of vertical angles.

Time

- 15–20 minutes

Materials/Software

- App: Cabri® Jr.
- AppVar: GL25
- Activity worksheet

Skills Needed

- drag an object
- animate an object

Classroom Management

- Students can work individually or in pairs depending on the number of calculators available.
- Use TI Connect™ software, TI-GRAPH LINK™ software, the TI-Navigator™ system, or unit-to-unit links to transfer GL25 to each calculator.

Notes

- Students can drag any labeled point except O.
- Students may answer the extension Questions 6 and 7 using reasoning or a drawing.

Answers

1. Check students' work.
2. $\angle AOB$ and $\angle COD$; $\angle BOC$ and $\angle AOD$
3. $\angle COD$; $m\angle AOB = m\angle COD$
4. their measures are equal
5. They are perpendicular.
6. 6 pairs
7. Answers may vary. Check students' work.

Name ______________________ Class ______________ Date ______________

Parallel Lines, Related Angles

Activity 31

FILES NEEDED: Cabri® Jr.
AppVar: GL31

Given: In GL31, $\overleftrightarrow{EF}$ intersects parallel lines $\overleftrightarrow{AB}$ and $\overleftrightarrow{CD}$ at points *P* and *R*.

Explore: relationships among angles formed by parallel lines and a transversal

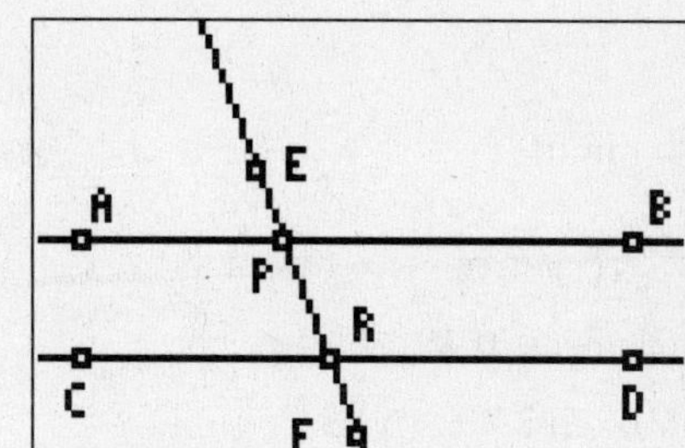

1. Complete the following. $\angle EPB$ and $\angle$? are corresponding angles. Copy the name of the second angle into the first column of the table below.

$m\angle EPB$				
$m\angle$ ____				

2. Install the screen-angle measures for the two angles. Drag point *E* horizontally to four locations. Record each pair of angle measures in the table above.

3. Study the data in the table. Complete the following conjecture.

   For parallel lines and a transversal, if two angles are corresponding angles, then ? .

4. Drag point *P* to a location above point *E* and complete the following statement. The two angles you named as corresponding angles in Question 1 are now ? angles.

5. Copy the angle name from Question 1 into the two blanks in the first column of the table below. Record the two current angle measures and their sum in the second column.

$m\angle EPB$				
$m\angle$ ____				
$m\angle EPB + m\angle$ ____				

6. Drag point *E* horizontally to three other locations. Record the angle measures and their sum in the table.

7. Study the data in the table. Write a conjecture based on your data.

Animation Option

In Questions 2 and 6 you dragged point *E*. Instead, animate point *E*. Then stop it at different locations to collect your data. (You can reset the animation by using Undo in the F1 menu.)

Parallel Lines, Related Angles

Teacher Notes

Activity Objective

Students use Cabri® Jr. to explore properties of corresponding and same-side interior angles.

Time

- 20–30 minutes

Materials/Software

- App: Cabri® Jr.
- AppVar: GL31
- Activity worksheet

Skills Needed

- install a measure
- drag an object
- animate an object

Classroom Management

- Students can work individually or in pairs depending on the number of calculators available.
- Use TI Connect™ software, TI-GRAPH LINK™ software, the TI-Navigator™ system, or unit-to-unit links to transfer GL31 to each calculator.

Notes

- In F1, select Open and then press ENTER to see the AppVar list.
- Students can drag points E and P. All other points are fixed, and line $\overleftrightarrow{AB}$ remains parallel to line $\overleftrightarrow{CD}$.

Error Prevention

- To install a screen-angle measure, be sure to choose a pre-existing point, which will flash when selected. If you do not select a pre-existing point, Cabri® Jr. will construct a new point which may not produce the desired results.

Answers

1–2. Check students' work.

3. they have equal measures

4. same-side interior

5–6. Check students' work.

7. Answers may vary. Sample: If two angles are same-side interior angles, they are supplementary.

Name ______________________ Class ______________ Date ______________

Exterior Angle of a Triangle

Activity 32

FILES NEEDED: Cabri® Jr.
AppVar: GL33

Given: In GL33, $\triangle BCD$ has exterior angle $\angle ABD$.

Explore: the measures of exterior angles

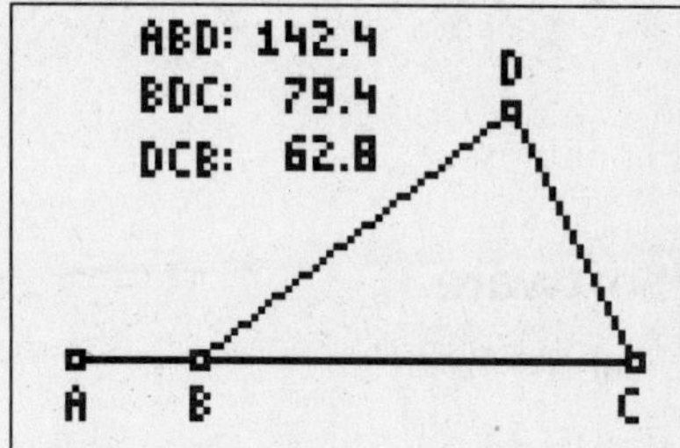

1. Drag point B to three different locations along $\overline{AC}$. Find $m\angle ABD$, $m\angle BDC$, and $m\angle DCB$ for each location. Collect your data in the first three blank columns of the table below.

$m\angle ABD$						
$m\angle BDC$						
$m\angle DCB$						

2. Drag point D vertically to three new locations. Find three more sets of values for $m\angle ABD$, $m\angle BDC$, and $m\angle DCB$. Record your data in the last three columns of the table above.

3. Study the data in the table. Complete the following conjecture about the exterior angle $\angle ABD$ and its two remote interior angles, $\angle BDC$ and $\angle DCB$.

   If $\angle ABD$ is an exterior angle of $\triangle BCD$, then $m\angle ABD$ = _?_ .

4. Generalize your conjecture from Question 3.

   The measure of an exterior angle of a triangle is _?_ .

Extension

5. How are $m\angle CBD$ and $m\angle ABD$ related?

6. How are $m\angle CBD$, $m\angle BDC$, and $m\angle DCB$ related?

7. Use your answers to Questions 5 and 6 to prove that your conjecture in Question 3 is true.

Animation Option

In Question 1, instead of dragging point B, animate B and stop it at different locations to collect your data. In Question 2, you can animate point D to collect data. (You can reset the animation by using Undo in the F1 menu.)

Exterior Angle of a Triangle

Teacher Notes

Activity Objective

Students use Cabri® Jr. to explore the relationship between interior and exterior angles of a triangle.

Time

- 15–20 minutes

Materials/Software

- App: Cabri® Jr.
- AppVar: GL33
- Activity worksheet

Skills Needed

- drag an object
- animate an object

Classroom Management

- Students can work individually or in pairs depending on the number of calculators available.
- Use TI Connect™ software, TI-GRAPH LINK™ software, the TI-Navigator™ system, or unit-to-unit links to transfer GL33 to each calculator.

Notes

- Students can drag point B along $\overline{AC}$. Students can drag point D freely.

Answers

1–2. Check students' work.

3. $m\angle BDC + m\angle DCB$

4. equal to the sum of the measures of the two remote interior angles

5. $m\angle CBD + m\angle ABD = 180.$

6. $m\angle CBD + m\angle BDC + m\angle DCB = 180.$

7. By substitution, $m\angle CBD + m\angle ABD = m\angle CBD + m\angle BDC + m\angle DCB.$ Subtract $m\angle CBD$ from each side. $m\angle ABD = m\angle BDC + m\angle DCB$

Name ____________________ Class ____________ Date ____________

Exterior Angle Sums

Activity 33

FILES NEEDED: Cabri® Jr.
AppVars: GL34A, GL34B

Given: In GL34A, $\angle 1$, $\angle 2$, and $\angle 3$ are exterior angles of the triangle.

Explore: the sum of the measures of exterior angles of a triangle

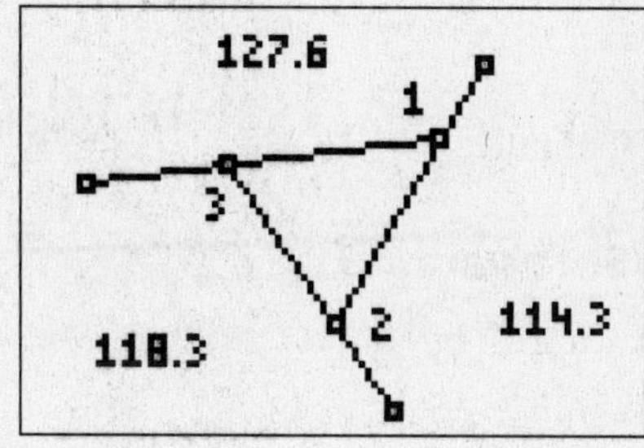

1. Drag any of the three vertices. Make four different triangles. For each triangle, find $m\angle 1$, $m\angle 2$, and $m\angle 3$. Record their measures and their sum in the table below.

$m\angle 1$				
$m\angle 2$				
$m\angle 3$				
$m\angle 1 + m\angle 2 + m\angle 3$				

2. Study the data in the table. Complete this conjecture about the sum of the measures of the exterior angles of a triangle.

The sum of the measures of the exterior angles of a triangle, one at each vertex, is _?_.

Given: In GL34B, $\angle 1$, $\angle 2$, $\angle 3$, and $\angle 4$ are exterior angles of the quadrilateral.

Explore: the sum of the measures of exterior angles of a quadrilateral

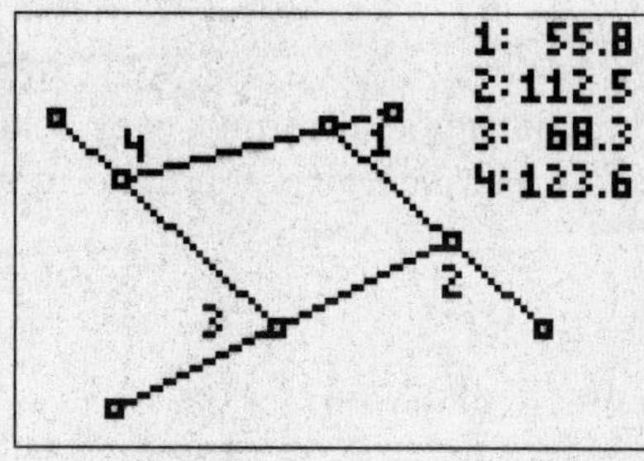

3. Open GL34B. Repeat Questions 1 and 2 for a quadrilateral. Record your data in the table below. Make a conjecture about the sum of the measures of the exterior angles of a quadrilateral.

$m\angle 1$				
$m\angle 2$				
$m\angle 3$				
$m\angle 4$				
$m\angle 1 + m\angle 2 + m\angle 3 + m\angle 4$				

Extension

4. Make a conjecture about the sum of the exterior angle measures of a pentagon.

Exterior Angle Sums

Activity Objective

Students use Cabri® Jr. to explore the sum of the measures of exterior angles of polygons.

Time

- 20–30 minutes

Materials/Software

- App: Cabri® Jr.
- AppVars: GL34A, GL34B
- Activity worksheet

Skills Needed

- drag an object

Classroom Management

- Students can work individually or in pairs depending on the number of calculators available.
- Use TI Connect™ software, TI-GRAPH LINK™ software, the TI-Navigator™ system, or unit-to-unit links to transfer GL34A and GL34B to each calculator.

Error Prevention

- Be sure that students form convex quadrilaterals only. Using concave quadrilaterals will produce unpredictable results.

Answers

1. Check students' work.
2. 360
3. Check students' work. The sum of the measures of the exterior angles of a quadrilateral is 360.
4. The sum of the exterior angle measures of a pentagon is 360.

Name ____________________ Class __________ Date __________

Angle Bisectors in Triangles I

Activity 34

FILES NEEDED: Cabri® Jr.
AppVar: GL45A

Given: In GL45A, $\overline{AT}$ bisects $\angle BAC$.

Explore: angle bisectors in triangles

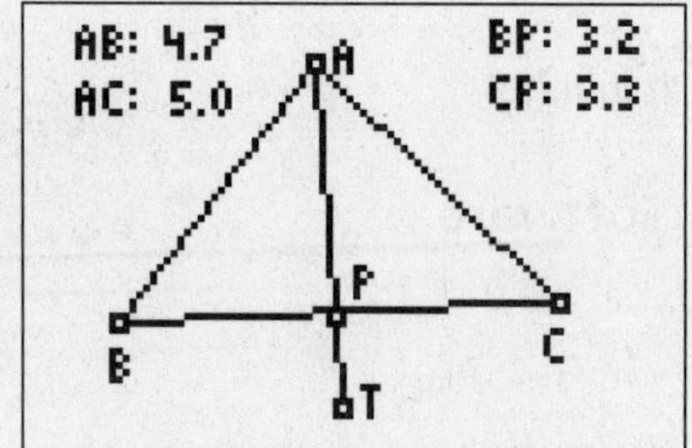

1. Drag point A, B, or C. Find four different isosceles triangles with $AB = AC$. For each triangle, record the lengths BP and CP in the table below.

BP				
CP				

2. Study the data in the table. Complete this conjecture about how lengths BP and CP are related.

   If the bisector of the vertex $\angle A$ of isosceles $\triangle ABC$ intersects the base $\overline{BC}$ in point P, then $BP = \underline{\ ?\ }$.

3. Generalize your conjecture from Question 2.

   The bisector of the vertex angle of an isosceles triangle $\underline{\ ?\ }$ the base of the triangle.

4. Install screen-angle measures for $\angle BPA$ and $\angle CPA$. Drag point A, B, or C. Find four different isosceles triangles with $AB = AC$. For each triangle, record $m\angle BPA$ and $m\angle CPA$ in the table below.

$m\angle BPA$				
$m\angle CPA$				

5. Study the data in the table. Complete this conjecture about $\angle BPA$ and $\angle CPA$.

   If the bisector of the vertex $\angle A$ of isosceles $\triangle ABC$ intersects the base $\overline{BC}$ in point P, then $\angle BPA$ and $\angle CPA$ $\underline{\ ?\ }$.

6. Generalize your conjecture from Question 5.

   The bisector of the vertex angle of an isosceles triangle $\underline{\ ?\ }$ the base of the triangle.

Extension

7. Combine your conjectures from Questions 3 and 6 into one statement.
8. Explain how to use GL45A to demonstrate the Isosceles Triangle Theorem.

Angle Bisectors in Triangles I

Teacher Notes

Activity Objective

Students use Cabri® Jr. to explore angle bisectors of isosceles triangles.

Time

- 15–20 minutes

Materials/Software

- App: Cabri® Jr.
- AppVar: GL45A
- Activity worksheet

Skills Needed

- drag an object
- install a measure

Classroom Management

- Students can work individually or in pairs depending on the number of calculators available.
- Use TI Connect™ software, TI-GRAPH LINK™ software, the TI-Navigator™ system, or unit-to-unit links to transfer GL45A to each calculator.

Notes

- In F1, select Open and then press ENTER to see the AppVar list.
- Students can drag only points A, B, and C. Points P and T are not draggable.
- Depending on the orientation of the triangle, it may not always be possible to match the lengths AB and AC exactly. Students can use values within one tenth of each other, or move B or C to reorient the base.

Answers

1. Check students' work.
2. CP
3. bisects
4. Check students' work.
5. are right angles
6. is perpendicular to
7. The bisector of the vertex angle of an isosceles triangle is the perpendicular bisector of the base of the triangle.
8. Answers may vary. Sample: Install screen-angle measures for $\angle B$ and $\angle C$. Find four different isosceles triangles with $AB = AC$. Record and study $m\angle B$ and $m\angle C$.

Name ______________________ Class ______________ Date ______________

Segment Bisectors in Triangles

Activity 35

FILES NEEDED: Cabri® Jr.
AppVar: GL45B

Given: In GL45B, point D bisects side $\overline{AC}$ of $\triangle ABC$.

Explore: segment bisectors in triangles

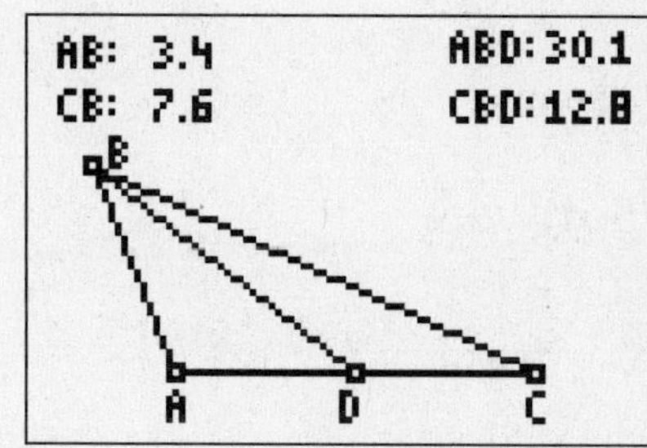

1. Drag point A, B, or C. Find four different isosceles triangles with $AB = CB$. For each triangle, record the angle measures indicated in the table below.

$m\angle ABD$				
$m\angle CBFD$				

2. Study the data in the table. Complete the following conjecture.

   If point D bisects the base $\overline{AC}$ of isosceles $\triangle ABC$, then $\overline{BD}$ __?__.

3. Drag point A, B, or C to get $AB = CB$ and $m\angle ABD$ as close to 45 as you can make it. What kind of triangle is $\triangle ABC$? Explain.

Extension

Make $\overline{AC}$ horizontal. Replace the screen measures AB and CB with DB and DC as shown at right.

4. Drag point B so that $\angle ABD$ and $\angle CBD$ are complementary. What kind of triangle is $\triangle ABC$? Explain.

5. Drag point B so that $\angle ABD$ and $\angle CBD$ are complementary in four different locations. In each location, what do you observe about DB and DC?

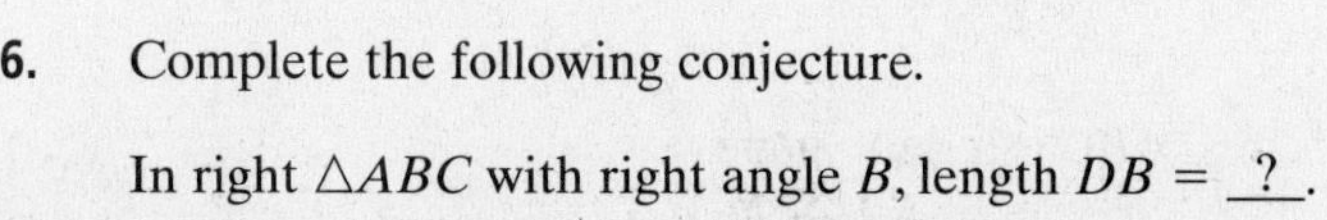

6. Complete the following conjecture.

   In right $\triangle ABC$ with right angle B, length DB = __?__.

7. Generalize your conjecture from Question 6.

   In a right triangle, the midpoint of the hypotenuse is __?__.

Segment Bisectors in Triangles

Teacher Notes

Activity Objective

Students use Cabri® Jr. to explore medians in isosceles triangles.

Time

- 15–20 minutes

Materials/Software

- App: Cabri® Jr.
- AppVar: **GL45B**
- Activity worksheet

Skills Needed

- drag an object

Classroom Management

- Students can work individually or in pairs depending on the number of calculators available.
- Use TI Connect™ software, TI-GRAPH LINK™ software, the TI-Navigator™ system, or unit-to-unit links to transfer **GL45B** to each calculator.

Notes

- If students cannot match the lengths AB and CB exactly, suggest that they move point A or C to reorient the base.
- You may wish to introduce the term *median* with this Activity.

Answers

1. Check students' work.

2. bisects $\angle ABC$

3. Right isosceles triangle; Since $AB = CB$, it is isosceles. Since $m\angle ABD = 45, m\angle CBD = 45$ and $m\angle ABC = 90$, so $\triangle ABC$ is a right triangle.

4. Right triangle; $m\angle ABD + m\angle CBD = 90°$, so $\angle ABC$ is a right angle.

5. They are equal.

6. $\frac{1}{2}$ the length of the hypotenuse

7. equidistant from the three vertices

Name ____________________ Class ____________ Date ____________

Triangle Midsegments

Activity 36

FILES NEEDED: Cabri® Jr.
AppVar: GL51

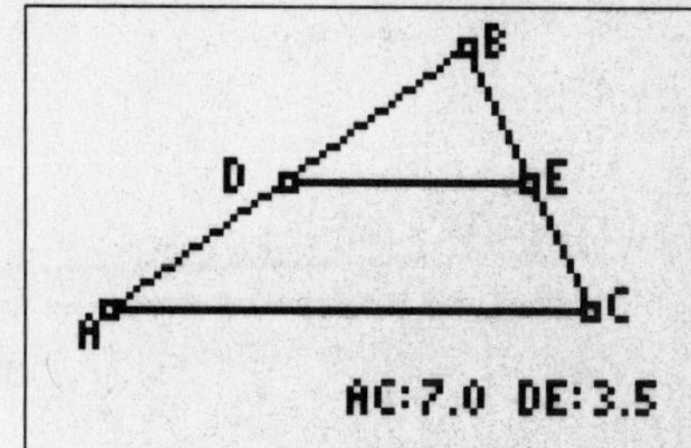

Given: In GL51, $\triangle ABC$ has sides $\overline{AB}$ and $\overline{BC}$ with midpoints D and E, respectively.

Explore: properties of a midsegment of a triangle

1. Drag point A to different locations. Collect data for the first three blank columns of the table below.

2. Drag point C to different locations. Collect data for the last three columns of the table.

Length DE						
Length AC						

3. Study the data in the table. Complete this conjecture about how lengths DE and AC are related.

   If D and E are midpoints of $\overline{AB}$ and $\overline{BC}$ in $\triangle ABC$, then DE = __?__.

4. Generalize your conjecture from Question 3.

   The length of a midsegment of a triangle is __?__.

Install the slope of $\overline{DE}$ beside $\overline{DE}$. Install the slope of $\overline{AC}$ beside $\overline{AC}$.

5. Drag point A to three different locations. Then drag point C to three different locations. For each location, collect data for the table below.

Slope of $\overline{DE}$						
Slope of $\overline{AC}$						

6. Study the data in the table. Make a conjecture about how the slopes of a midsegment of a triangle and the corresponding base are related.

Extension

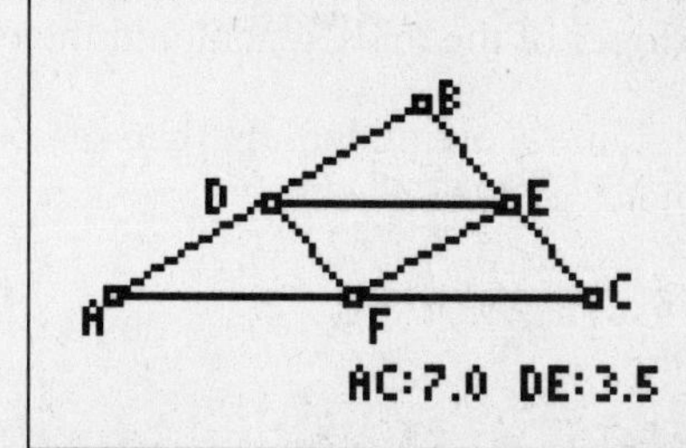

7. Construct the midpoint of $\overline{AC}$ and segments $\overline{DF}$ and $\overline{FE}$ to complete $\triangle DEF$. Make a conjecture about how the perimeters of $\triangle DEF$ and $\triangle ABC$ are related.

8. Install the two perimeters on the screen. Drag vertices to check your conjecture.

Triangle Midsegments

Teacher Notes

Activity Objective

Students use Cabri® Jr. to explore properties of triangle midsegments.

Time

- 20–30 minutes

Materials/Software

- App: Cabri® Jr.
- AppVar: GL51
- Activity worksheet

Skills Needed

- drag an object
- construct a midpoint
- install a measure
- draw a segment

Classroom Management

- Students can work individually or in pairs depending on the number of calculators available.
- Use TI Connect™ software, TI-GRAPH LINK™ software, the TI-Navigator™ system, or unit-to-unit links to transfer GL51 to each calculator.

Notes

- Students cannot drag points D and E because D and E are constructed as the midpoints of segments $\overline{AB}$ and $\overline{BC}$.
- Dragging point B will not cause changes in the lengths nor the slopes of $\overline{DE}$ or $\overline{AC}$. Students can investigate this property of $\triangle ABC$ as well.

Answers

1–2. Check students' work.

3. $\frac{1}{2}AC$

4. half the length of the corresponding base

5. Check students' work.

6. The slopes of the midsegment and the corresponding base are equal.

7. If D, E, and F are midpoints, then the perimeter of $\triangle DEF$ is half the perimeter of $\triangle ABC$.

8. Check students' work.

Name ______________________________ Class ________________ Date ________________

Perpendicular Bisectors in Triangles

Activity 37

FILES NEEDED: Cabri® Jr.
AppVar: GL53

Given: in GL53, $\triangle ABC$

Explore: perpendicular bisectors in triangles

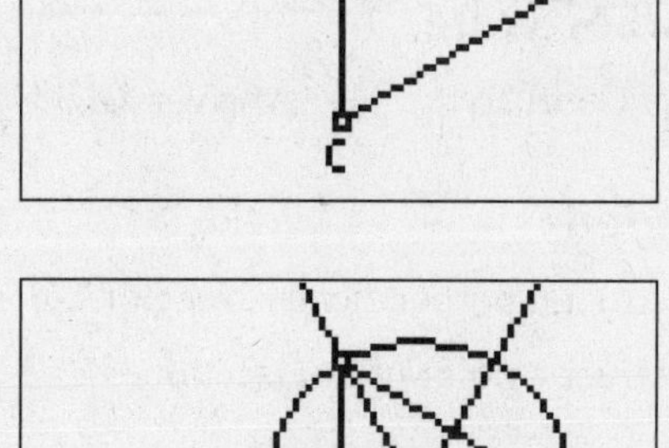

Step 1: For $\triangle ABC$, construct the perpendicular bisector of sides $\overline{AB}$ and $\overline{AC}$.

Step 2: Construct the point at the intersection of the two perpendicular bisectors. Name it P.

Step 3: Construct a circle centered at P. For the radius, join P to one vertex of the triangle.

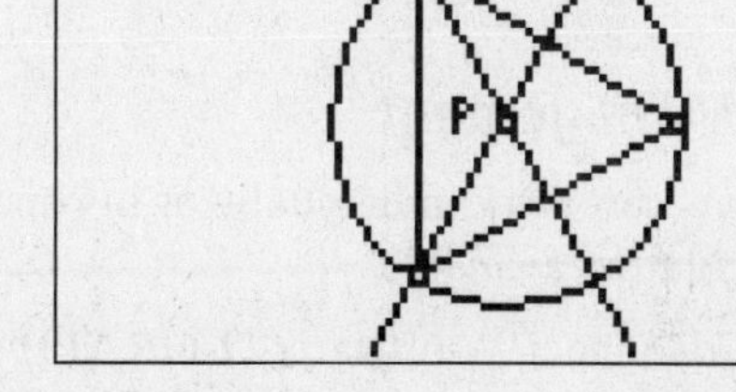

1. Why can you conclude that $PA = PB = PC$?

2. What can you conclude about the circle you drew in Step 3?

3. The circle is the *circumcircle* of the triangle and point P is the *circumcenter*. Explain why *circumcircle* is a fitting name for this circle.

4. Drag vertex A toward side $\overline{BC}$ until the circumcenter lies on $\overline{BC}$. Estimate $m\angle BAC$.

5. Drag vertex A closer to $\overline{BC}$. Does the circumcenter remain within the triangle? Does the circle still pass through all three vertices? Justify your answer.

6. Describe the type of triangle to complete each conjecture.

If the circumcenter is inside the triangle, the triangle is _?_.
If the circumcenter is outside the triangle, the triangle is _?_.

7. If your conjectures in Question 6 are true, what can you conclude about the triangle if the circumcenter lies on the triangle?

Extension

8. As vertex A gets very close to $\overline{BC}$, what happens to P? To the circumcircle?

9. If A lies on $\overline{BC}$, what happens to the perpendicular bisectors? To P? To the circle?

Animation Option

Animate vertex A and watch it pass back and forth through $\overline{BC}$.
Note how the perpendicular bisectors and the size of the circle change.

Perpendicular Bisectors in Triangles

Teacher Notes

Activity Objective

Students use Cabri® Jr. to explore properties of perpendicular bisectors of triangles.

Time

- 20–30 minutes

Materials/Software

- App: Cabri Jr.
- AppVar: GL53
- Activity worksheet

Skills Needed

- construct a perpendicular bisector
- construct a point of intersection
- construct a circle
- drag an object
- animate an object

Classroom Management

- Students can work individually or in pairs depending on the number of calculators available.
- Use TI Connect™ software, TI-GRAPH LINK™ software, the TI-Navigator™ system, or unit-to-unit links to transfer GL53 to each calculator.

Notes

- $\triangle ABC$ is constructed so that point A can be animated.

Error Prevention

- When students attach the radius to one of the vertices of $\triangle ABC$, they need to be sure the vertex is flashing before they press ENTER.

Answers

1. A point on the perpendicular bisector of a segment is equidistant from the endpoints.
2. It contains all three vertices of the triangle.
3. *circum-* (Latin) means around.
4. $m\angle BAC \approx 90°$.
5. no; yes, same as Question 1
6. acute; obtuse
7. The triangle is a right triangle.
8. P gets farther and farther away from $\overline{BC}$. The circumcircle becomes larger and larger.
9. They become parallel through P at infinity and the circle becomes a line.

Name ______________________ Class ______________ Date ______________

Inequalities in Triangles

Activity 38

FILES NEEDED: Cabri® Jr.
AppVar: GL55

Given: in GL55, $\triangle ABC$

Explore: the side lengths and angle measures of triangles

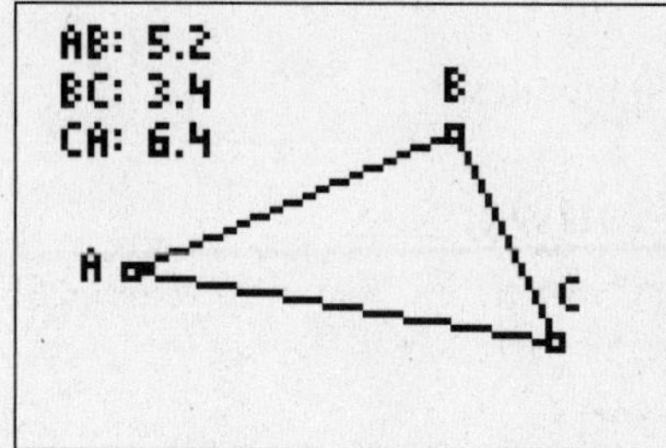

1. Drag any vertex of triangle $\triangle ABC$ to form four different triangles. Record lengths AB, BC, and CA for each triangle in the table below.

Length AB				
Length BC				
Length CA				
Sum of two smallest lengths				

2. In each column of the table, ring the greatest length. Write the sum of the two smallest lengths in the bottom row.

3. Study the data in the table. Make a conjecture about an inequality relationship between the sum of the lengths of the two shorter sides of a triangle and the length of the longest side.

4. Generalize your conjecture from Question 3 to a conjecture about the sum of the lengths of *any* two sides of a triangle and the length of the third side.

5. Assume that the conjecture from Question 3 is true. Then give a convincing argument that your new conjecture from Question 4 is true, or find a counterexample to show that it is false.

Extension

6. Install the screen-angle measures for $\angle A$, $\angle B$, and $\angle C$. Drag any vertex of triangle $\triangle ABC$ to form four different triangles. In the table below, record $m\angle A$, $m\angle B$, and, $m\angle C$ for each triangle, along with the length of the side opposite each angle.

$m\angle A$, length BC								
$m\angle B$, length CA								
$m\angle C$, length AB								

7. For each triangle ring the greatest angle measure and greatest length. Box the smallest angle measure and smallest length. Make a conjecture.

Inequalities in Triangles

Teacher Notes

Activity Objective

Students use Cabri® Jr. to explore inequalities in triangles.

Time

- 20–30 minutes

Materials/Software

- App: Cabri Jr.
- AppVar: GL55
- Activity worksheet

Skills Needed

- install a measure
- drag an object

Classroom Management

- Students can work individually or in pairs depending on the number of calculators available.
- Use TI Connect™ software, TI-GRAPH LINK™ software, the TI-Navigator™ system, or unit-to-unit links to transfer GL55 to each calculator.

Error Prevention

- Some students may form an equilateral triangle. In this case, tell them that they can consider any side to be the longest side.
- It is possible to form triangles for which the sum of the lengths of two sides of the triangle appear to be *equal to* the length of the third side. This is due to the limited precision of the screen measurements.

Answers

1–2, 6. Check students' work.

3. The sum of the lengths of the two shorter sides is greater than the length of the longest side.

4. The sum of the lengths of any two sides of a triangle is greater than the length of the third side.

5. Sample: Question 4 conjecture must be true. The sum for any two sides is no less than the sum for the two shorter sides. Also, any side is no longer than the longest side.

7. The largest angle and longest side are opposite each other as are the smallest angle and shortest side.

Name ______________________ Class ______________ Date ______________

Parallel Segments in Triangles

Activity 39

FILES NEEDED: Cabri® Jr.
AppVar: GL85A

Given: In GL85A, $\overline{DE}$ is parallel to side $\overline{AC}$ of $\triangle ABC$ with point D on $\overline{AB}$ and point E on $\overline{BC}$.

Explore: the ratios of the lengths of the divided sides

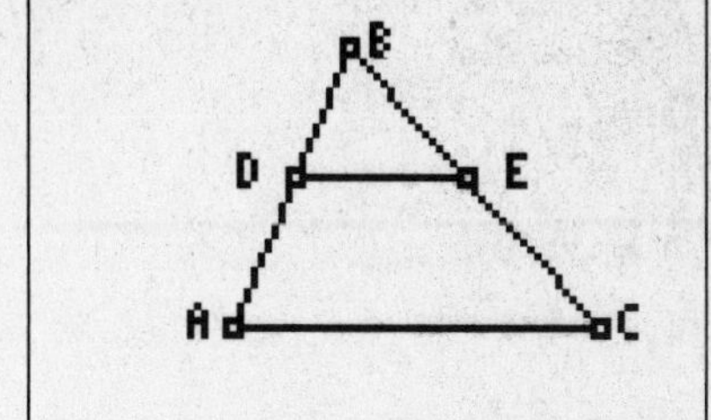

1. Install lengths AD, DB, CE, and EB on your screen. Drag point D to four different locations on $\overline{AB}$. For each location record the four lengths in the table below.

Length AD				
Length DB				
Length CE				
Length EB				
$\frac{AD}{DB}$				
$\frac{CE}{EB}$				

2. For each column in the table, find the ratios $\frac{AD}{DB}$ and $\frac{CE}{EB}$ to the nearest tenth. Record the values in the last two rows of the table.

3. Study the data in the table. Complete the following conjecture about the relationship between $\frac{AD}{DB}$ and $\frac{CE}{EB}$.

   If $\overline{DE}$ is parallel to side $\overline{AC}$ of $\triangle ABC$ with D on $\overline{AB}$ and E on $\overline{BC}$, then __?__.

4. Generalize your conjecture from Question 3.

   If a line is parallel to one side of a triangle and intersects the other two sides, then __?__.

Extension

5. Drag point C until $\triangle ABC$ is an isosceles triangle with $AB = CB$. What kind of triangle is $\triangle DBE$? Explain your answer.

6. Draw $\overline{GH}$ parallel to $\overline{AC}$ of $\triangle ABC$ with point G on $\overline{AD}$ and point H on $\overline{EC}$. Make a conjecture about the ratio $AG:GD:DB$.

Parallel Segments in Triangles

Teacher Notes

Activity Objective

Students use Cabri® Jr. to explore the properties of parallel segments in triangles.

Time

- 20–30 minutes

Materials/Software

- App: Cabri® Jr.
- AppVar: GL85A
- Activity worksheet

Skills Needed

- install a measure
- drag an object

Classroom Management

- Students can work individually or in pairs depending on the number of calculators available.
- Use TI Connect™ software, TI-GRAPH LINK™ software, the TI-Navigator™ system, or unit-to-unit links to transfer GL85A to each calculator.

Answers

1-2. Check students' work.

3. $\frac{AD}{DB} = \frac{CE}{EB}$

4. the line divides the two sides proportionally

5. $\triangle DBE$ is isosceles; If $\frac{AD}{DB} = \frac{CE}{EB}$, then $\frac{AD + DB}{DB} = \frac{CE + EB}{EB}$, or $\frac{AB}{DB} = \frac{CB}{EB}$. When $AB = CB$, it follows that $DB = EB$, so $\triangle DBE$ is isosceles.

6. $AG:GD:DB = CH:HE:EB$

Name ______________________ Class ______________ Date ______________

Angle Bisectors in Triangles II

Activity 40

FILES NEEDED: Cabri® Jr.
AppVar: GL85B

Given: In GL85B, $\overleftrightarrow{AD}$ bisects $\angle CAB$ of $\triangle ABC$ and intersects side $\overline{CB}$ in point D.

Explore: the ratio of the lengths of the divided side

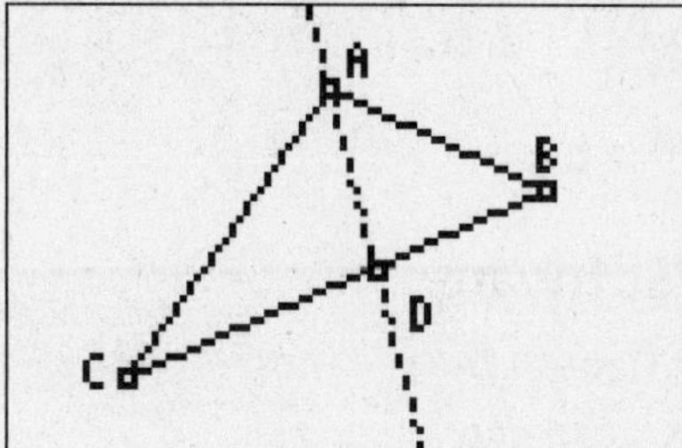

1. Install lengths CA, AB, CD, and DB on your screen. Drag point B to form four different triangles. For each triangle record the four lengths in the table below.

Length CA				
Length AB				
Length CD				
Length DB				
$\frac{CA}{AB}$				
$\frac{CD}{DB}$				

2. For each column in the table, find the ratios $\frac{CA}{AB}$ and $\frac{CD}{DB}$ to the nearest tenth. Record the values in the last two rows of the table.

3. Study the data in the table. Complete the following conjecture about the relationship between $\frac{CD}{DB}$ and $\frac{CA}{AB}$.

If $\overleftrightarrow{AD}$ bisects $\angle CAB$ of $\triangle ABC$ and intersects side $\overline{CB}$ in point D, then __?__.

4. Generalize your conjecture from Question 3.

A line that bisects an angle of a triangle divides the side opposite that angle into two segments whose lengths are proportional to __?__.

Extension

5. Drag point B to make $CD = DB$. What kind of triangle is $\triangle ABC$? Explain your answer.

6. Explain how, without measuring any angle, you could locate a point E on $\overline{AC}$ so that $\overleftrightarrow{BE}$ bisects $\angle ABC$.

Angle Bisectors in Triangles II

Teacher Notes

Activity Objective

Students use Cabri® Jr. to explore the properties of angles bisectors in triangles.

Time

- 20–30 minutes

Materials/Software

- App: Cabri® Jr.
- AppVar: GL85B
- Activity worksheet

Skills Needed

- install a measure
- drag an object

Classroom Management

- Students can work individually or in pairs depending on the number of calculators available.
- Use TI Connect™ software, TI-GRAPH LINK™ software, the TI-Navigator™ system, or unit-to-unit links to transfer GL85B to each calculator.

Answers

1-2. Check students' work.

3. $\frac{CD}{DB} = \frac{CA}{AB}$

4. the lengths of the sides of the angle

5. isosceles; If $CD = DB$, then $\frac{CA}{AB} = 1$, and $CA = AB$.

6. If $\overleftrightarrow{BE}$ bisects $\angle ABC$, $\frac{AE}{EC}$ must equal $\frac{BA}{BC}$. Install lengths AE, EC, BA, and BC. Locate E so that $AE = EC \cdot \frac{BA}{BC}$. Then $\frac{AE}{EC} = \frac{BA}{BC}$.

Name ______________________ Class ______________ Date ________

Triangles and Circles

Activity 41

FILES NEEDED: Cabri® Jr.
AppVars: GL72A, GL72B

Given: In GL72A, $\triangle ABC$ has $\overline{AB}$ as a diameter of the circle and point C on the circle.

Explore: $\triangle ABC$

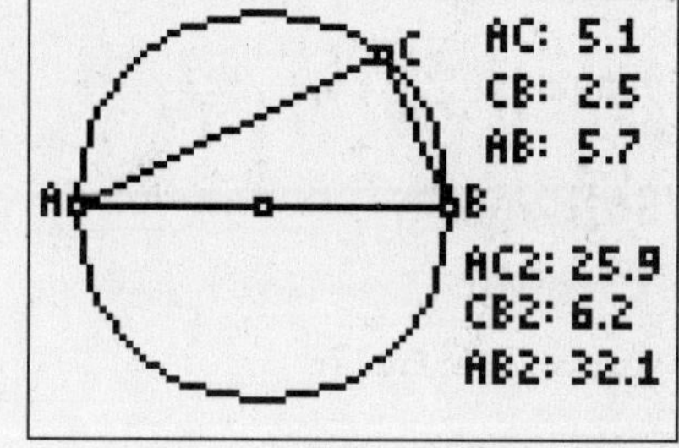

1. Drag point C to four different locations on the circle. (See Animation Option below.) Record the values of AC^2, CB^2, and AB^2 in the table below.

AC^2				
CB^2				
AB^2				

2. Study the table. How are AC^2, CB^2, and AB^2 related?

3. Recall the Converse of the Pythagorean Theorem. Use it to help you complete the following conjecture.

If $\overline{AB}$ is the diameter of a circle and point C lies on the circle, then $\triangle ABC$ is _?_.

4. Generalize your conjecture from Question 3 to a conjecture about a triangle inscribed in a circle with one of its sides a diameter of the circle.

Extension

In GL72B, point P is the center of the circle. $\triangle ABC$ has point C on the circle and $\overline{AB}$ as a diameter.

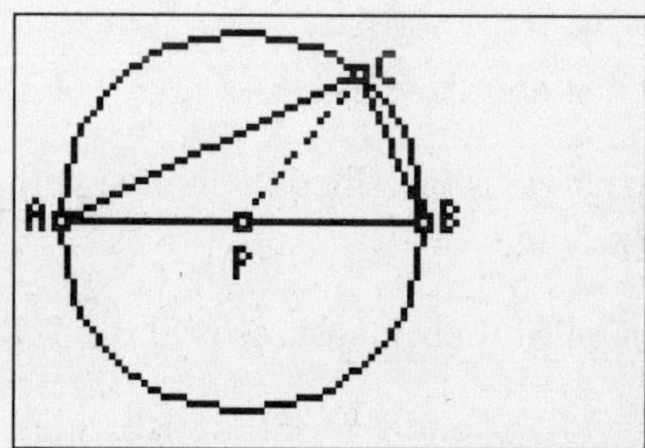

5. Install the screen measures of $\angle ACP$ and $\angle BCP$. Drag point C around the circle. What do you notice about $m\angle ACP$ and $m\angle BCP$?

6. What is true about $\overline{PA}$, $\overline{PB}$, and $\overline{PC}$? What kind of triangles are $\triangle APC$ and $\triangle BPC$?

7. Write a proof of your conjecture in Question 3. (*Hint:* Use your answers from Question 6.)

Animation Option

In Question 1, animate point C and stop it at different locations to collect your data.

Triangles and Circles

Teacher Notes

Activity Objective

Students use Cabri® Jr. to explore properties of triangles inscribed in circles.

Time

- 20–30 minutes

Materials/Software

- App: Cabri® Jr.
- AppVar: GL72A, GL72B
- Activity worksheet

Skills Needed

- drag an object
- install a measure

Classroom Management

- Students can work individually or in pairs depending on the number of calculators available.
- Use TI Connect™ software, TI-GRAPH LINK™ software, the TI-Navigator™ system, or unit-to-unit links to transfer GL72A and GL72B to each calculator.
- For Question 5, suggest that students collect data in a table.

Notes

- AC2: 25.9 on the calculator screen means "$AC^2 = 25.9$."

Answers

1. Check students' work.
2. $AC^2 + CB^2 = AB^2$
3. a right triangle
4. If a triangle is inscribed in a circle with one of its sides a diameter, then it is a right triangle.
5. The sum of their measures is about 90.
6. They are congruent; isosceles.
7. Since $\triangle APC$ and $\triangle BPC$ are isosceles, $m\angle A = m\angle ACP$ and $m\angle B = m\angle BCP$. $m\angle A + m\angle B + m\angle ACB = 180$, so $m\angle ACP + m\angle BCP + m\angle ACB = 180$. But $m\angle ACP + m\angle BCP = m\angle ACB$, so $m\angle ACB + m\angle ACB = 180$, or $2\, m\angle ACB = 180$. Thus, $m\angle ACB = 90$ and $\triangle ABC$ is a right triangle.

Name ______________________ Class ______________ Date ______________

Reflections

Activity 42

FILE NEEDED: Cabri® Jr.

1. Follow the steps to draw, animate, and demonstrate a dynamic image and its equally-dynamic reflection image.

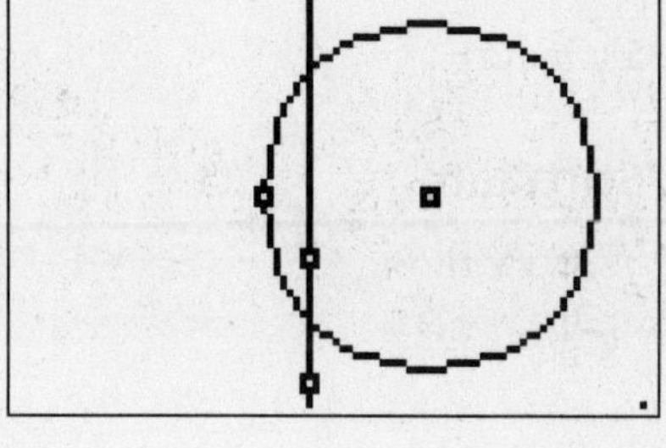

Step 1: Draw a vertical line.

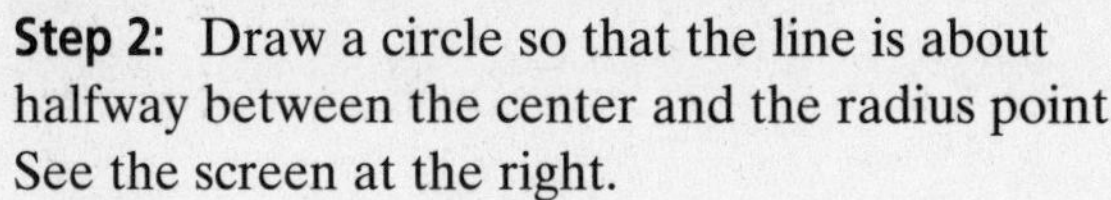

Step 2: Draw a circle so that the line is about halfway between the center and the radius point. See the screen at the right.

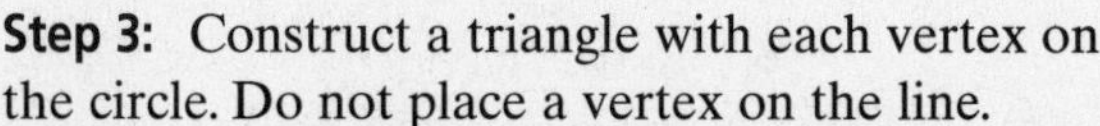

Step 3: Construct a triangle with each vertex on the circle. Do not place a vertex on the line.

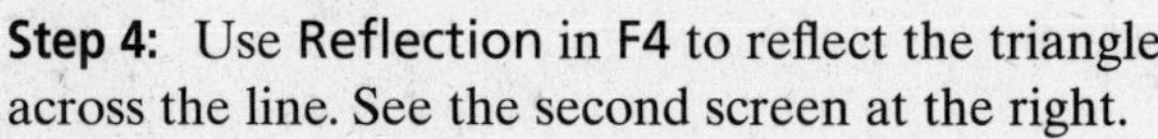

Step 4: Use **Reflection** in **F4** to reflect the triangle across the line. See the second screen at the right.

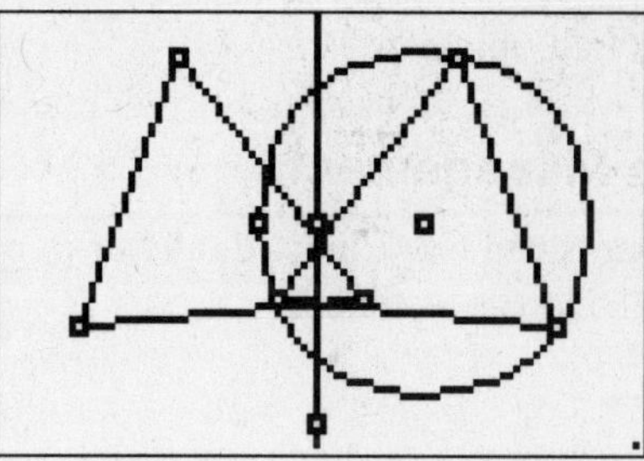

Step 5: Hide the circle, its center and radius points, and the vertical line and its defining points.

Step 6: Choose one vertex on the original triangle. Animate it. After you see how the triangle and its reflection change, animate another vertex. Finally animate the third vertex.

2. Stop the animation at a place of your choosing. Sketch the symmetric figure you see. Draw the line of symmetry in your sketch.

Extension

3. To make the animation more interesting, follow Steps 7 and 8.

Step 7: Construct a segment between each vertex and its reflection.

Step 8: Animate the vertices, one at a time, until all three are moving.

4. How is the vertical line from Step 1 related to the three segments you constructed in Step 7? Explain.

5. How are the three segments you constructed in Step 7 related to each other? Explain.

Reflections

Teacher Notes

Activity Objective

Students use Cabri® Jr. to explore reflected images.

Time

- 25–35 minutes

Materials/Software

- App: Cabri® Jr.
- Activity worksheet

Skills Needed

- draw a line
- animate a point
- reflect an object
- hide an object
- construct a triangle
- draw a circle
- construct a segment

Classroom Management

- Students can work individually or in pairs depending on the number of calculators available.

Notes

- The relative positions of the line and circle impacts the appearance of the construction. If the line sits on the edge of the circle, the animated points will meet but not cross over.
- Animate in F1 activates one point at a time. Students can generate different images by varying the time between animating points and the activation sequence.
- Students who complete the activity quickly should try their own variations.

Answers

1-3. Check students' work.

4. It is the perpendicular bisector of each segment by the definition of reflection.

5. They are parallel. Lines perpendicular to the same line are parallel to each other.

Name ________________ Class ________________ Date ________________

Area of a Triangle

Activity 43

FILES NEEDED: Cabri® Jr.
AppVars: GL71A, GL71B

Given: In GL71A, $\overleftrightarrow{DE} \parallel \overleftrightarrow{AC}$ and $\overline{BP} \perp \overleftrightarrow{AC}$.
$\triangle ABC$ has base length AC and height BP.

Explore: area of $\triangle ABC$

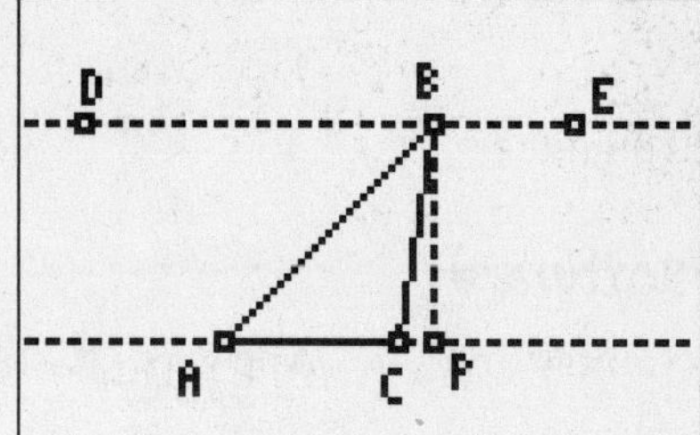

1. Install the screen measures BP and area of $\triangle ABC$. Predict what will happen to each screen measure as you drag point B along $\overleftrightarrow{DE}$.

2. Justify each prediction. Then test your predictions by dragging B along $\overleftrightarrow{DE}$.

Before doing Question 2, save your GL71A from Question 1 as PIC1.

In GL71B at the right, $\triangle ABC$ is the same triangle as the one shown above. In this case, however, $\overleftrightarrow{EF} \parallel \overline{AB}$.

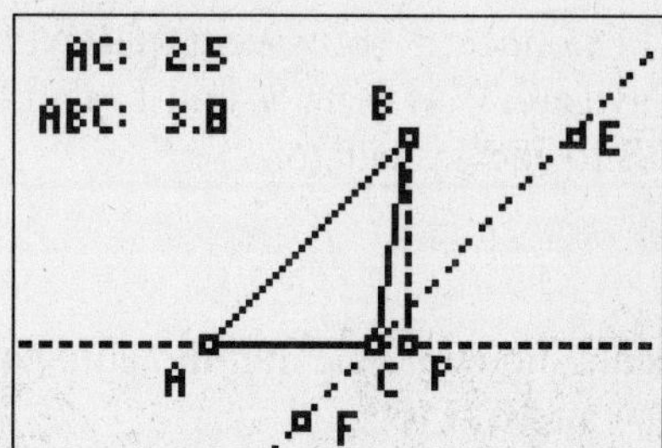

3. Predict what will happen to the screen measures AC and area of $\triangle ABC$ as you drag C along $\overleftrightarrow{EF}$.

4. Justify each prediction. Then test your predictions by dragging C along $\overleftrightarrow{EF}$.

5. For each of three locations of C, predict the value of BP. Then test your predictions by installing the screen measure BP. If your predictions are correct, explain why.

Extension

Recall the screen that you saved as PIC1. Replace the screen measures for BP with the measures for AB and BC. Also, install the screen measure for the perimeter of $\triangle ABC$. Note that you now have four measures on the screen.

6. Drag point B along $\overleftrightarrow{DE}$. Describe $\triangle ABC$ for large values of the perimeter and for small values of the perimeter.

7. Drag B to find the smallest value of the perimeter. What type of triangle does $\triangle ABC$ appear to be? Give a convincing argument why $\triangle ABC$ must be this type of triangle.

Area of a Triangle

Teacher Notes

Activity Objective

Students use Cabri® Jr. to explore the relationship between the base and height and the area of a triangle.

Time

- 15–20 minutes

Materials/Software

- App: Cabri® Jr.
- AppVars: GL71A, GL71B
- Activity worksheet

Skills Needed

- drag an object
- install a measure

Classroom Management

- Use TI Connect™ software, TI-GRAPH LINK™ software, the TI-Navigator™ system, or unit-to-unit links to transfer GL71A and GL71B to each calculator.

Notes

- Students should notice that the initial area of $\triangle ABC$ is the same in GL71A and GL71B.

Answers

1. The height and the area will not change.
2. $\overleftrightarrow{DE}$ and $\overleftrightarrow{AC}$ are parallel, so BP will not change. The area does not change because it depends on base and height, which do not change.
3. AC will change. The area will stay the same.
4. AC increases as C moves away from A toward E. AC decreases as C moves closer to A in the direction of F. Area stays the same because 1) base $\overline{AB}$ does not change and 2) the height to $\overline{AB}$ stays the same as parallel lines remain a constant distance apart.
5. Check students' work. $BP = \frac{2 \cdot \text{Area } \triangle ABC}{AC}$ (both numerator and denominator shown on screen).
6. $\triangle ABC$ is obtuse for large perimeters and acute for small perimeters.
7. Isosceles; Answers may vary. Sample: For every non-isosceles triangle, there is a second triangle congruent to it. These two triangles determine two locations of B. The triangle for each location of B between these two points has a smaller perimeter. Thus the smallest perimeter must occur where $BA = BC$.

Name ______________________ Class ______________ Date ______________

Areas of Rhombuses and Kites

Activity 44

FILES NEEDED: Cabri® Jr.
AppVars: GL74A, GL74B

Given: In GL74A, *RHMB* is a rhombus with diagonals $\overline{RM}$ and $\overline{HB}$.

Explore: the area of a rhombus

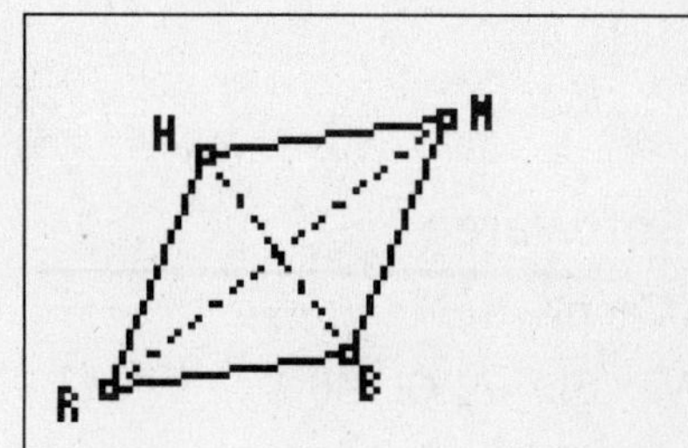

1. Install the screen measures for *RM*, *HB*, and the area of rhombus *RHMB*. Drag point *H* or *B*. For four different rhombuses, record each measure in the table below.

RM				
HB				
Area				

2. Study the data in the table. Make a conjecture that relates the area of rhombus *RHMB* to the lengths of diagonals $\overline{RM}$ and $\overline{HB}$.

3. Reword your conjecture from Question 2 to relate the area of *any* rhombus to the lengths of its diagonals.

In GL74B, *KITE* is a kite with diagonals $\overline{KT}$ and $\overline{IE}$.

4. Install the screen measures for *KT*, *IE*, and the area of *KITE*. Drag point *K, I,* or *T*. For four different kites, record each measure in the table below.

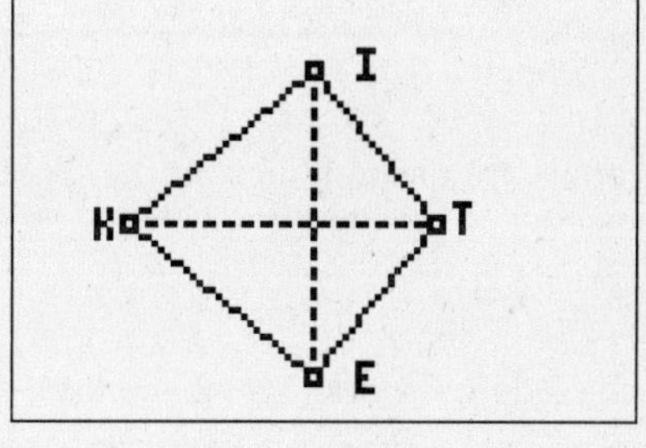

KT				
IE				
Area				

5. Study the data in the table. Make a conjecture that relates the area of any kite to the lengths of its diagonals.

Extension

6. A square is a rhombus. Reword your conjecture from Question 3 to make a conjecture about the area of a square.

7. A rectangle has congruent diagonals, just like a square. In your conjecture from Question 6, replace "square" with "rectangle" to form a new conjecture. Is this conjecture true? Give a convincing argument.

8. Prove that your conjectures from Questions 3 and 5 are true.

Areas of Rhombuses and Kites

Teacher Notes

Activity Objective

Students use Cabri® Jr. to explore areas of rhombuses and kites.

Time

- 20–30 minutes

Materials/Software

- App: Cabri® Jr.
- AppVar: GL74A, GL74B
- Activity worksheet

Skills Needed

- install a measure
- drag an object

Classroom Management

- Students can work individually or in pairs depending on the number of calculators available.
- Use TI Connect™ software, TI-GRAPH LINK™ software, the TI-Navigator™ system, or unit-to-unit links to transfer GL74A and GL74B to each calculator.

Notes

- Diagonals are dashed lines on the screen.

Answers

1. Check students' work.

2. Area of $RHMB = \frac{1}{2}(RM)(HB)$

3. The area of a rhombus is one half the product of the lengths of its diagonals.

4. Check students' work.

5. The area of a kite is one half the product of the lengths of its diagonals.

6. The area of a square is one half the square of the length of a diagonal.

7. The area of a rectangle is half the square of the length of a diagonal. False. You can rotate the diagonals to form a variety of rectangles that have different areas.

8. Proofs may vary. Sample (referring to rhombus *RHMB* on the Activity page with diagonals intersecting at point *P*): Area $RHMB$ = Area $\triangle RHM$ + Area $\triangle MBR$ = $\frac{1}{2}(RM)(HP) + \frac{1}{2}(RM)(PB) = \frac{1}{2}(RM)(HP + PB) = \frac{1}{2}(RM)(HB)$.

Name ____________________ Class ____________ Date ____________

Perimeters and Areas of Squares

Activity 45

FILES NEEDED: Cabri® Jr.
AppVar: GL86

Given: In GL86, $ABCD$ and $AEFG$ are squares.

Explore: the ratios of side lengths, perimeters, and areas of squares

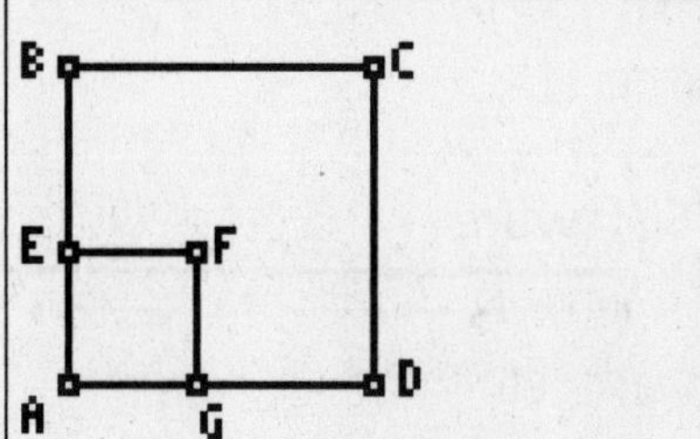

1. On your screen, install the lengths AB and AE, and the perimeters and areas of $ABCD$ and $AEFG$. Drag points D and G to form four different pairs of squares. Record all six measures for each pair in the table below.

Length AB				
Length AE				
Perimeter of $ABCD$				
Perimeter of $AEFG$				
Area of $ABCD$				
Area of $AEFG$				

2. Use data from the table above. For each column, find and record the ratios shown in the table below to the nearest tenth.

$\frac{AB}{AE}$				
$\frac{\text{Perimeter of } ABCD}{\text{Perimeter of } AEFG}$				
$\frac{(AB)^2}{(AE)^2}$				
$\frac{\text{Area of } ABCD}{\text{Area of } AEFG}$				

3. Study the data in the table. Make two conjectures about the ratios shown in the table.

4. Generalize your conjectures from Question 3 to apply to all squares. (*Hint:* The perimeters of any two squares are proportional to . . .)

Animation Option

Animate points D and G to form the different squares.

Perimeters and Areas of Squares

Teacher Notes

Activity Objective

Students use Cabri® Jr. to explore the perimeters and areas of squares.

Time

- 20–30 minutes

Materials/Software

- App: Cabri® Jr.
- AppVar: GL86
- Activity worksheet

Skills Needed

- install a measure
- drag an object

Classroom Management

- Students can work individually or in pairs depending on the number of calculators available.
- Use TI Connect™ software, TI-GRAPH LINK™ software, the TI-Navigator™ system, or unit-to-unit links to transfer GL86 to each calculator.

Notes

- Ask students if they think the same relationships will apply to rectangles or triangles. Have them explain their answers.

Answers

1-2. Check students' work.

3. $\frac{AB}{AE} = \frac{\text{Perimeter of } ABCD}{\text{Perimeter of } AEFG}$.

$\frac{(AB)^2}{(AE)^2} = \frac{\text{Area of } ABCD}{\text{Area of } AEFG}$.

4. The perimeters of two squares are proportional to the lengths of their sides. The areas of two squares are proportional to the squares of the lengths of their sides.

Name ______________________ Class ______________ Date ______________

Inscribed Angles

Activity 46

FILES NEEDED: Cabri® Jr.
AppVars: GL113A, GL113B

Given: In GL113A, points B, C, D and E, are on a circle with center A. $\overline{DE}$ (only endpoints shown) is a diameter of the circle.

Explore: inscribed angles

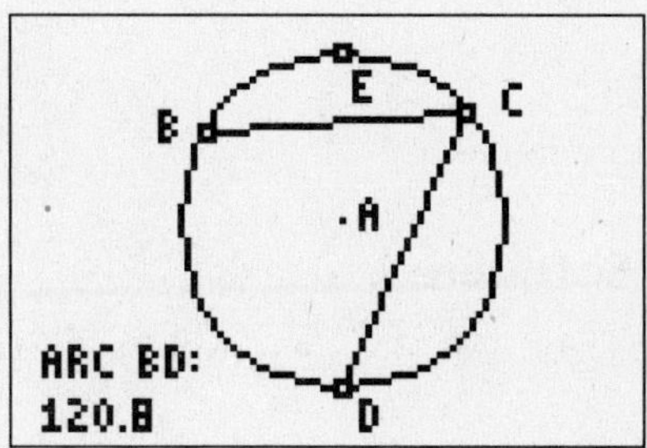

1 Install the screen measure of $\angle BCD$.

2. Drag both points B and C along the circle to form four different angles BCD. Keep B to the left and C to the right of diameter $\overline{DE}$. Record the measures listed in the table below.

$m\angle BCD$				
$m\widehat{BD}$				

3. Study the data in the table. Make a conjecture about the relationship between $m\angle BCD$ and $m\widehat{BD}$.

4. Generalize your conjecture from Question 3 to a conjecture about the measure of any inscribed angle and its intercepted arc.

5. Predict what will happen if you drag point C along the circle without moving B? Justify your prediction. Then test it by dragging C.

6. $\widehat{BD}$ is intercepted by both inscribed $\angle BCD$ and central angle $\angle BAD$. Make a conjecture relating the measures of an inscribed angle and a central angle that intercept the same arc. Install $m\angle BAD$ and check your conjecture.

Extension

Open GL113B. $\overline{RT}$ is a diameter of the circle and P is a point on the circle.

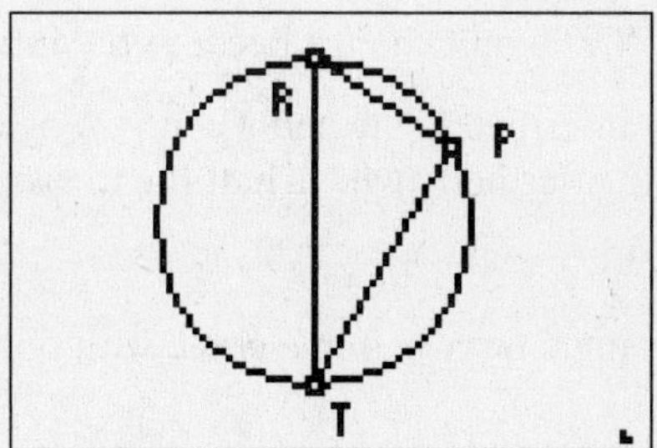

7. Install the screen measure of $\angle RPT$. Drag point P. (See Animation Option below.) What do you observe?

8. Make a conjecture about a triangle inscribed in a circle and with one side a diameter.

Animation Option

Animate point P along the circle.

Inscribed Angles

Teacher Notes

Activity Objective

Students use Cabri® Jr. to explore the relationship between inscribed angles and their intercepted arcs.

Time

- 10–15 minutes

Materials/Software

- App: Cabri® Jr.
- AppVars: GL113A, GL113B
- Activity worksheet

Skills Needed

- install a measure
- drag an object
- animate an object

Classroom Management

- Students can work individually or in pairs depending on the number of calculators available.
- Use TI Connect™ software, TI-GRAPH LINK™ software, the TI-Navigator™ system, or unit-to-unit links to transfer AppVars GL113A and GL113B to each calculator.

Error Prevention

- Make sure students do not form intercepted arcs greater than 180º. Cabri Jr. will not generate angle measures greater than 180.

Answers

1–2. Check students' work.

3. If $\overparen{BD}$ is the arc intercepted by inscribed $\angle BCD$, then $m\angle BCD = \frac{1}{2}m\overparen{BD}$.

4. The measure of an inscribed angle is equal to half the measure of its intercepted arc.

5. $m\angle BCD$ won't change because the intercepted arc does not change.

6. If an inscribed angle and a central angle intercept the same arc, the measure of the inscribed angle is half the measure of the central angle.

7. $m\angle RPT = 90$

8. A triangle inscribed in a circle with one side a diameter is a right triangle.

Name ______________________ Class ______________ Date ______________

Parallels and Perpendiculars

Activity 47

FILES NEEDED: Transformation Graphing App Program: A2L22

A2L22 graphs the linear equation or function $y = mx + b$ as Y1 = AX + B.

In this activity you will explore the relationships between the equation of a line and the slopes of lines parallel and perpendicular to its graph.

1. Run A2L22 to see a fixed line (two points marked) and line Y1 = AX + B with A = 0 and B = 5. From looking at the screen, how would you change A to make the two lines parallel?

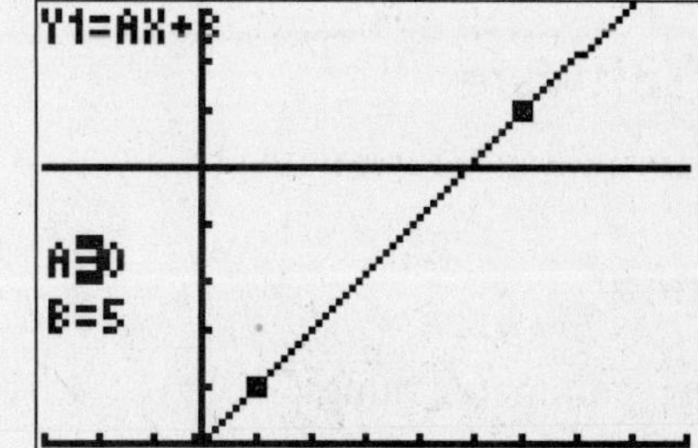

2. Change the value of A until the lines appear to be parallel. Record this value in the table below.

3. Change the value of A until the lines appear to be perpendicular. Record this value in the table.

4. Use TRACE to find the coordinates of the two marked points on the fixed line. Record these in the table. Then use the slope formula $m = \frac{y_2 - y_1}{x_2 - x_1}$ to fill in the last column.

	Slope of parallel line	Slope of perpendicular line	Coordinates of marked points	Slope of fixed line
Questions 2–4				
Question 5				
Question 6				

5. Switch from Plot1 to Plot2. Repeat Questions 2–4 for the new fixed line.

Plot1 Plot2 Plot3
Y1 = AX+B
Y2 =
Y3 =
Y4 =
Y5 =
Y6 =
Y7 =

6. Switch from Plot2 to Plot3. Repeat Questions 2–4. (*Hint:* When you look for the perpendicular line, first change the value of B to −4.)

7. Make a conjecture about the slopes of parallel lines; about the slopes of perpendicular lines.

8. Test your conjectures. Let Y1 = 3X + 7. Write equations for Y2 and Y3 so that their graphs are parallel and perpendicular, respectively, to the graph of Y1. Graph the lines in a square window.

Extension

9. Write equations of a line parallel $2x + 5y = 7$ and of a line perpendicular to $2x + 5y = 7$.

Parallels and Perpendiculars

Teacher Notes

Activity Objective

Students use the Transformation Graphing App to explore the relationships between the equation of a line and the slopes of lines parallel and perpendicular to its graph.

Time

- 20–25 minutes

Materials/Software

- Transformation Graphing App
- Program: A2L22
- Activity worksheet

Skills Needed

- change parameter values
- select and deselect plots

Notes

- Reminder: As noted on p. vi, each Activity page assumes that you activate the appropriate App at the start of the activity.
- Be certain that students understand which line is the graph of Y1 = AX + B.
- Remind students that they can enter parameter values A and B directly.
- Discuss with students how to deselect and select a plot in either the Y= or STAT PLOT screens.

Answers

1. Increase the value of A.

2–6. See table below.

	Slope of parallel line	Slope of perpendicular line	Coordinates of marked points	Slope of fixed line
Questions 2–4	1	−1	(1, 1) (6, 6)	1
Question 5	0.5	−2	(2, 1) (6, 3)	$\frac{1}{2}$
Question 6	−0.25	4	(0, 3) (4, 2) (8, 1)	$-\frac{1}{4}$

7. The slopes of parallel lines are equal; the slopes of perpendicular lines are opposite reciprocals of each other.

8. Check students' work. The parallel line must have slope 3. The perpendicular line must have slope $-\frac{1}{3}$.

9. Check students' work. The parallel line must have slope $-\frac{2}{5}$. The perpendicular line must have slope $\frac{5}{2}$.

Name ____________________ Class ____________ Date ____________

Visualizing Linear Models

Activity 48

FILES NEEDED: Transformation Graphing App
Programs: A2L24A, A2L24B, A2L24C

When you run each program, the startup screen shows the linear equation $y = mx + b$ graphed as Y1 = AX + B, and the startup values for A and B.

For each scatter plot, manipulate the line Y1 to find the best trend line that you can to model the data. Write the linear equation and answer the related questions.

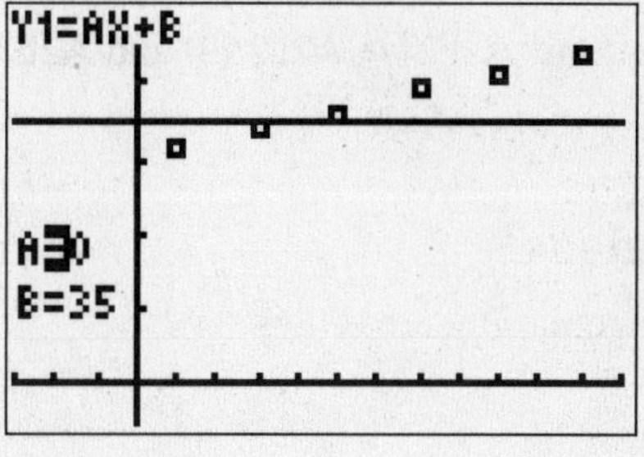

1. For text Exercise 12, run A2L24A and find a trend line to model the data. (*Hint:* Find a reasonable slope; then the y-intercept.) Use ENTER to change parameter values.
2. For your model, find the European equivalent of U.S. size 8.
3. Explain how you can use your model to find the U.S. equivalent of a European size.

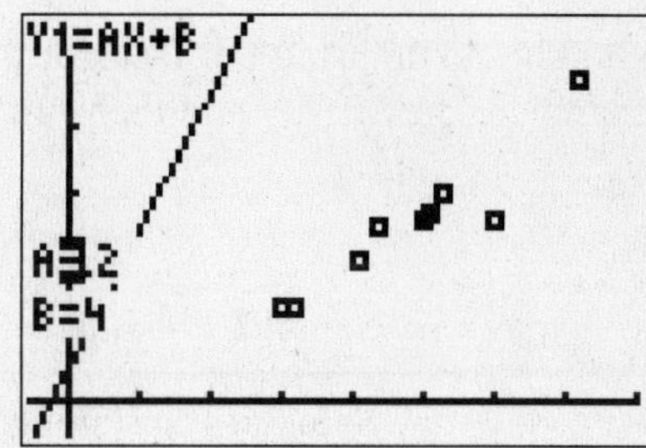

4. For text Exercise 20, run A2L24B and find a trend line to model the data. Round the value for the slope to hundredths.
5. How much fat would you expect a 330-Calorie hamburger to have?
6. What is the meaning of the y-intercept? What value would you expect for it?

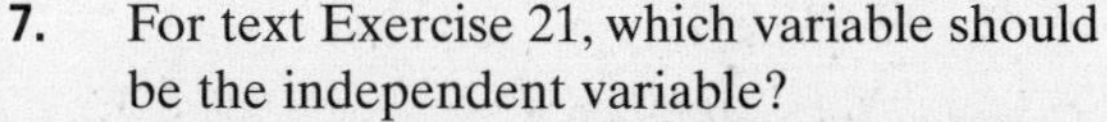

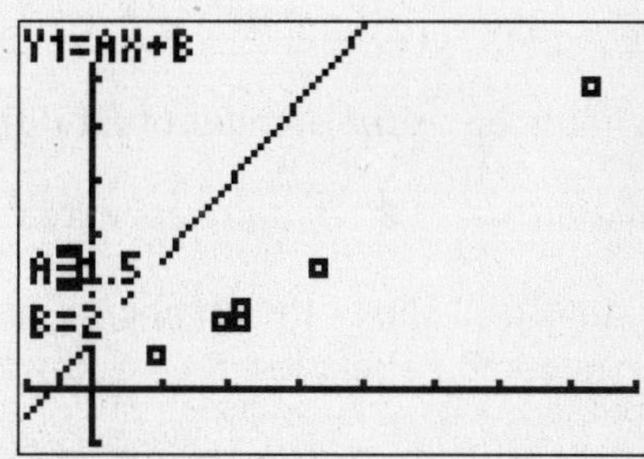

7. For text Exercise 21, which variable should be the independent variable?
8. Run A2L24C. Find a trend line to model the data. Find A and B to tenths.
9. When the data were collected, the population of Oregon was approximately 3 million. Use your model to estimate how many licensed drivers there were in Oregon.
10. What is the meaning of the y-intercept? What value would you expect for it?
11. Which point represents Florida? Ignore that point and find your best model for the remaining data. How do your two models differ and why?
12. Use your second model to estimate the number of licensed drivers in Oregon. In which estimate, this one or the one in Exercise 9, would you have more confidence? Explain.

Visualizing Linear Models

Teacher Notes

Activity Objective

Students use the Transformation Graphing App to find a trend line to model data.

Time

- 20–25 minutes

Materials/Software

- Transformation Graphing App
- Programs: A2L24A, A2L24B, and A2L24C
- Activity worksheet

Skills Needed

- change parameter values

Notes

- Questions 11 and 12 show how an outlier affects trend lines.
- Students should uninstall the Transformation Graphing App when they complete the activity and then run DEFAULT.

Answers

1. Answers may vary. Sample: $y = 1.3x + 30$

2. size 40

3. Answers may vary. Sample: Trace along the trend line and find the x-value (U.S. size) for a given y-value (European size).

4. Answers may vary. Sample: $y = 0.07x - 9$

5. about 14 g

6. The y-intercept is the amount of fat if there were no Calories. It should be 0.

7. population

8. Answers may vary. Sample: $y = 0.8x - 0.5$

9. 1.9 million

10. The y-intercept shows the number of licensed drivers you would expect if the state had population zero. You would expect this value to be 0.

11. The point at the far upper right of the screen. Answers may vary. Sample: $y = 0.7x$. The Florida data increase the slope of the trend line.

12. 2.1 million; check students' work.

Name ____________________ Class ____________ Date ____________

Absolute Value Translations I

Activity 49

FILES NEEDED: Transformation Graphing App
Program: A2L25

A2L25 graphs the absolute value function $y = |mx + b| + c$ as Y1 = abs(AX + B) + C. In this activity you will study the relationship between the vertex of a graph and the values of m, b, and c.

1. Run A2L25. The startup graph has A = 1, B = 0, and C = 0. Write the equation for this absolute value function. What are the coordinates of its vertex?

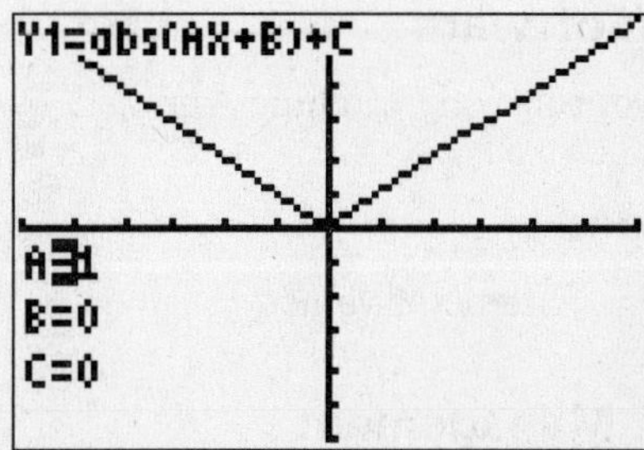

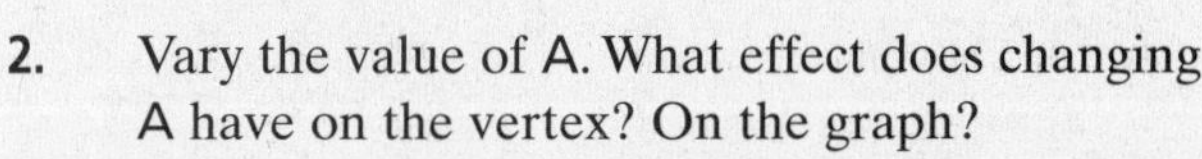

2. Vary the value of A. What effect does changing A have on the vertex? On the graph?

3. Set A = 1 and B = 0, and vary the value of C. What effect does changing C have on the vertex?

4. Set A = 1 and C = 0, and vary the value of B. What effect does changing B have on the vertex?

5. When A = 1, predict the coordinates of the vertex for each pair of B and C values given below. Test your predictions.

B = 2, C = 3 B = 4, C = −2 B = −1, C = 3 B = −4, C = −1

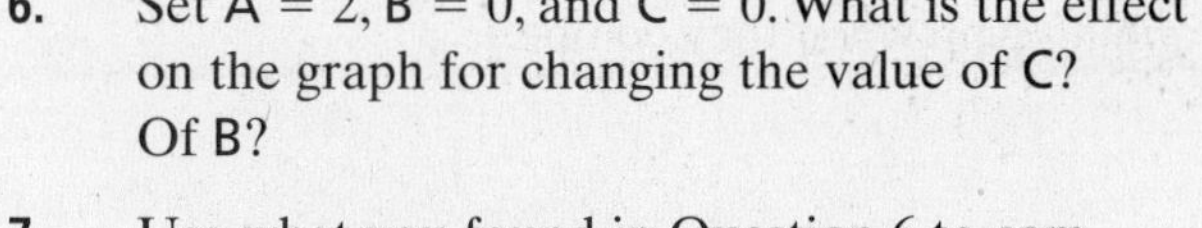

6. Set A = 2, B = 0, and C = 0. What is the effect on the graph for changing the value of C? Of B?

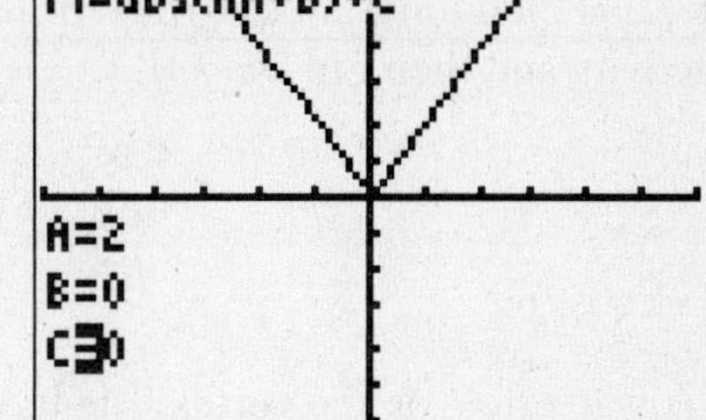

7. Use what you found in Question 6 to complete this conjecture.

When A = 2, for every increase by 1 in B, the vertex moves _?_ unit(s).

Set A = 2, B = 0, and C = −2. Test your conjecture. Revise your conjecture as needed.

8. Set A = 4, B = 0, and C = 1. Vary the B values and complete the conjecture.

When A = 4, for every increase by 1 in B, the vertex moves _?_ unit(s).

9. Generalize your findings from the questions above.

In terms of m, b, and c, the coordinates of the vertex of the graph of the absolute value function $y = |mx + b| + c$ are _?_.

10. Test your conjecture with the following A, B, and C values. Revise your conjecture as needed.

A = 2, B = 2, C = 4 A = 2, B = 6, C = −2 A = 2, B = −6, C = 2

A = 4, B = 8, C = −2 A = 4, B = −10, C = −3 A = 4, B = −16, C = 2

Absolute Value Translations I

Teacher Notes

Activity Objective

Students use the Transformation Graphing App to explore how changing parameter values affects the graph of an absolute value function.

Time

- 30–40 minutes

Materials/Software

- Transformation Graphing App
- Program: A2L25
- Activity worksheet

Skills Needed

- change parameter values

Classroom Management

- This activity can be used as a teacher demonstration.

Notes

- A2L25 uses the form Y1 = |AX + B| + C rather than $y = |mx + b| + c$. Students have to interpret "A" as "m."
- Remind students that they need to use the (-) key, not the subtraction key, to enter a negative number.
- Students should uninstall the Transformation Graphing App when they complete the activity and then run DEFAULT.

Answers

1. $y = |x|$; $(0, 0)$

2. There is no effect on the vertex. The larger the absolute value of A the steeper the sides of the V.

3. Changing C moves the vertex up or down.

4. Changing B moves the vertex left or right.

5. $(-2, 3)$; $(-4, -2)$; $(1, 3)$; $(4, -1)$

6. Changing C moves the vertex up or down. Changing B moves the vertex left or right.

7. -0.5

8. -0.25

9. $\left(-\frac{b}{m}, c\right)$

10. Check students' work.

Name ____________________ Class ____________ Date ____________

Absolute Value Translations II

Activity 50

FILES NEEDED: Transformation Graphing App
Program: A2L26

A2L26 graphs the absolute value function $y = |x - b| + c$ (sometimes written $y = |x - h| + k$) as Y1 = abs(X − B) + C. In this activity you will study how B and C are related to vertical and horizontal translations of the graph.

1. Run A2L26 and complete each sentence. The startup graph has B = _?_ and C = _?_. It shows the *parent* absolute value function $y =$ _?_.

2. Vary the value of C. What is the effect on the graph of $y = |x| + c$ when you increase the value of C? Decrease the value of C?

3. Predict the y-coordinate of the vertex for each function below. Test each prediction by adjusting the value of C on your screen.

   $y = |x| + 3$ $\quad$ $y = |x| - 2$ $\quad$ $y = |x| + \frac{3}{2}$

4. Write the absolute value function for each graph shown. Check your answers by drawing each graph with your calculator.

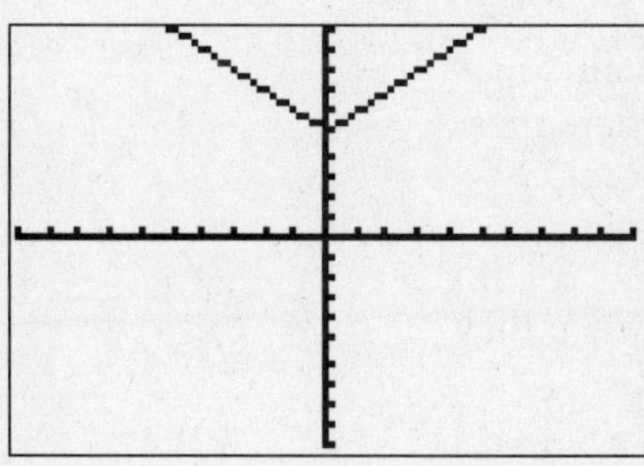

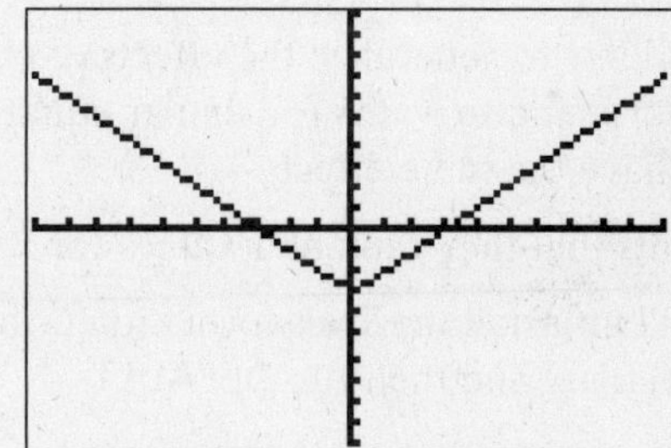

5. Set C = 0. Vary the value of B. What is the effect on the graph of $y = |x - b|$ when you increase the value of B? Decrease the value of B?

6. Write the function equations $y = |x - b|$ for B = 4 and for B = −2. Predict where the vertex will be located in each graph. Test each prediction by adjusting the value of B on your screen.

7. Complete each conjecture.

   $y = |x + 4|$ translates the graph of $y = |x|$ by _?_ units to the _?_.

   $y = |x - 5|$ translates the graph of $y = |x|$ by _?_ units _?_.

8. Generalize your conjectures from Questions 2 and 7.

   $y = |x - b| + c$ translates the graph of $y = |x|$ by _?_ and by _?_.

9. Test your conjecture. Predict the location of the vertex of the graph of each function. Test your predictions by adjusting the screen values of B and C.

   $y = |x - 3| + 2$ $\quad$ $y = |x - 4| - 2$ $\quad$ $y = |x + 1| - 3$ $\quad$ $y = |x + 5| + 4$

Absolute Value Translations II

Teacher Notes

Activity Objective

Students use the Transformation Graphing App to investigate horizontal and vertical translations of the absolute value function.

Time

- 25–30 minutes

Materials/Software

- Transformation Graphing App
- Program: A2L26
- Activity worksheet

Skills Needed

- change parameter values

Notes

- Discuss the concept of a parent function before this activity. Here students are investigating the translations of the parent function $y = |x|$.
- The Transformation Graphing App uses only A, B, C, and D as parameter names, so $y = |x - h| + k$ appears in the form Y1 = abs(X − B) + C.
- Encourage students to generalize the effects of changes in B and C in this equation form and to watch for similar equation forms where these two constants have the same effects.
- Remind students that they can enter values for B and C directly.
- Students should uninstall the Transformation Graphing App when they complete the activity and then run DEFAULT.

Answers

1. 0; 0; $|x|$

2. An increase translates the graph up. A decrease translates the graph down.

3. $3, -2, \frac{3}{2}$

4. $y = |x| + 5; y = |x| - 3$

5. As B increases the graph moves to the right. As B decreases the graph moves to the left.

6. $y = |x - 4|, y = |x + 2|; (4, 0), (-2, 0)$

7. 4, left; 5, to the right

8. b units right for positive values of b and $|b|$ units left for negative values of b; c units up for positive values of c and $|c|$ units down for negative values of c.

9. $(3, 2); (4, -2); (-1, -3); (-5, 4)$

Name ______________________ Class ______________ Date ______________

General Inequality Systems

Activity 51

FILES NEEDED: Inequality Graphing App
Program: A2L33A, A2L33B, A2L33C, A2L33D

Sometimes an inequality in a system of inequalities is nonlinear. In this activity you practice graphing systems that include such forms.

For each system, describe the graph and boundaries in words. Then draw the graph. Run the indicated program to check your answers.

1. $y \geq |x - 3| - 2$
$y \leq -3x + 5$

Run A2L33A to check. Note that you have to insert the correct inequality symbols here, and in each of Questions 2–4.

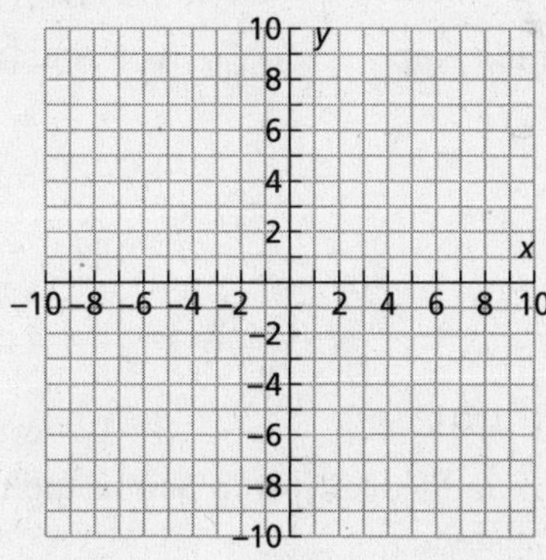

2. $y > 2|x + 1| - 2$
$y \leq 0.5x + 4$

Run A2L33B to check.

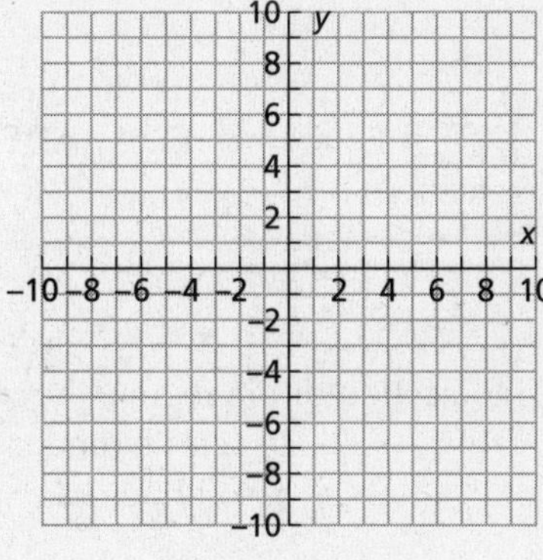

3. $y < -3x + 8$
$y \geq 2x - 3$
$y \leq 5x + 5$

Run A2L33C to check.

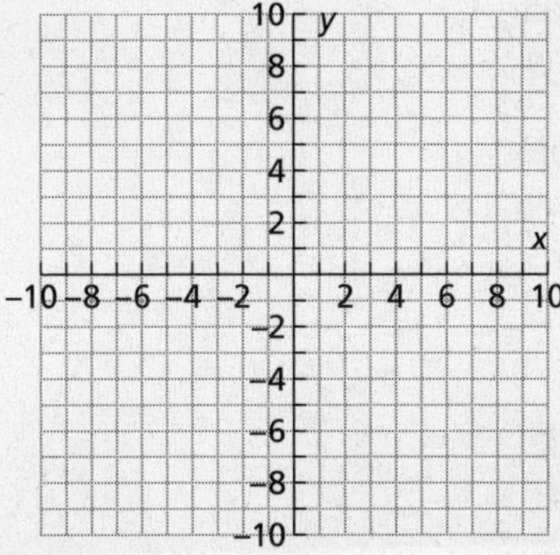

4. $y \leq x^2 - 4$
$y \geq -|x| + 3$
$y < 5$

Run A2L33D to check.

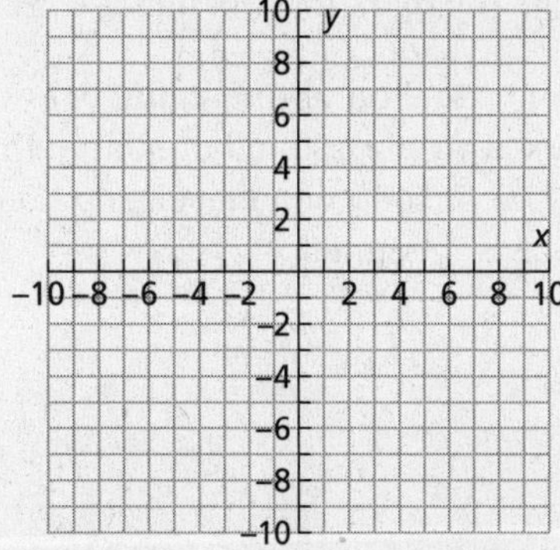

General Inequality Systems

Teacher Notes

Activity Objective

Students use the Inequality Graphing App to graph systems that include nonlinear inequalities and systems with more than two inequalities.

Time

- 10–15 minutes

Materials/Software

- Inequality Graphing App
- Activity worksheet
- Programs: A2L33A, A2L33B, A2L33C

Skills Needed

- insert inequality sign

Classroom Management

- Students can work individually or in pairs depending on the number of calculators available.

Notes

- To see the graph of a system even better in Questions 1–4, remind students to press ALPHA F2 to open the SHADES menu and then make the correct selection.

Answers

1. The graph is the intersection of the region above the graph of $y = |x - 3| - 2$ [vertex at $(3, -2)$, solid boundary], and the region below the graph of $y = -3x + 5$, solid boundary.

2. The graph is the intersection of the region above the graph of $y = 2|x + 1| - 2$ [vertex at $(-1, -2)$, dashed boundary] and the region below the graph of $y = 0.5x + 4$, solid boundary.

3. The graph is the intersection of the region below the line $y = -3x + 8$ (dashed boundary), the region above the line $y = 2x - 3$ (solid boundary), and the region below the line $y = 5x + 5$ (solid boundary).

4. The graph is the intersection of the region below the parabola $y = x^2 - 4$ [vertex at $(0, -4)$, solid boundary], the region above the graph of $y = -|x| + 3$ (solid boundary), and the region below the graph of $y = 5$ (dashed boundary).

Name ______________________ Class ______________ Date ______________

Vertex Principle

Activity 52

FILES NEEDED: Transformation Graphing App
Program: A2L34A

In this activity you will use the Transformation Graphing App to support the Vertex Principle of Linear Programming, which states:

If there is a maximum or a minimum value of the linear objective function, it occurs at one or more vertices of the feasible region.

A2L34A graphs the linear equation or function $y = mx + b$ as Y1 = AX + B.

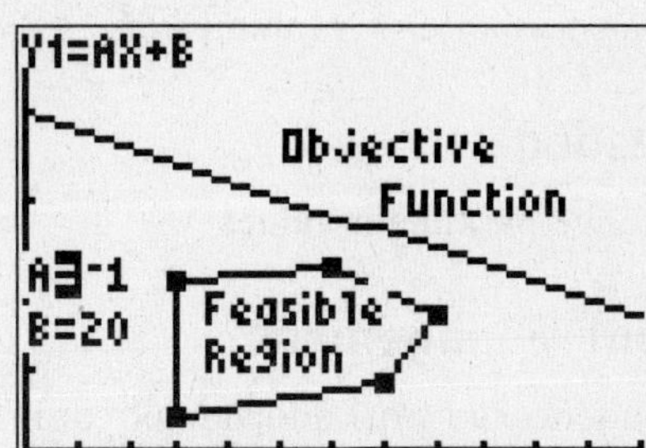

1. Write the equation for the linear function graphed in the startup screen. Write it in slope-intercept form and also in linear objective-function form, $P = ax + by$.
2. For the linear objective function $P = ax + by$, the idea is to find the values of feasible x and y that give the extreme (largest or smallest) value of P. Thus the objective function represents a family of functions determined by all possible values of P. Write the objective function in slope-intercept form.
3. Use your response to Question 2. What is the slope of the graphs of the functions in the family of the objective function? Explain why the slope is constant.
4. What does your response to Question 3 tell you about the graphs of the functions in the family of the objective function?
5. Use your response to Question 2. How are the y-intercepts of the objective function family related to P?

Run A2L34A. The line represents one member of the objective function family whose slope is the constant A; in this case -1.

6. Vary the value of B. Give a plausible argument why the extreme value of the objective function occurs at a vertex of the feasible region.
7. Change the value of the slope, A, to see a member of a different objective function family. Vary B as in Question 6. Is your argument from Question 6 still plausible?
8. Repeat Questions 6 and 7 for Plot2 and Plot3.

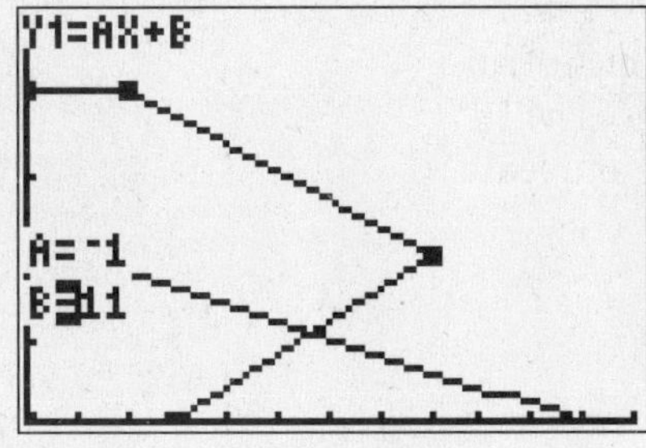

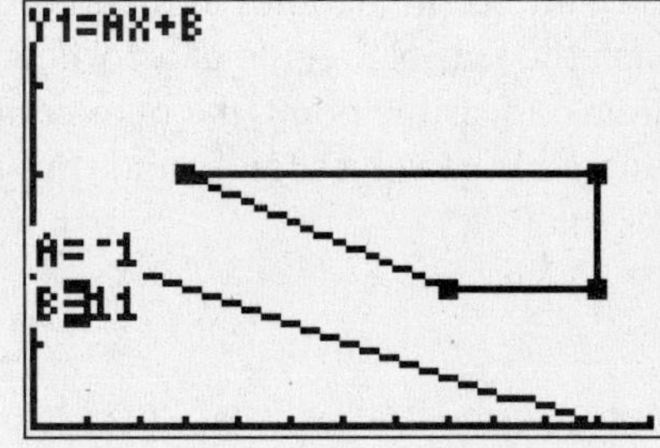

Vertex Principle

Activity Objective

Students use the Transformation Graphing App to help understand the Vertex Principle of Linear Programming.

Time

- 30–35 minutes

Materials/Software

- Transformation Graphing App
- Program: A2L34A
- Activity worksheet

Skills Needed

- change parameter values
- switch plots

Classroom Management

- Students can work individually or in pairs depending on the number of calculators available.

Notes

- When you switch to Plot2 or Plot3, the parameter values do not change so the initial screens may be different from what you see below Question 8.
- You may want to review with students how to deselect and select a plot in either the Y = or STAT PLOT screens.

Answers

1. $y = -x + 20; 20 = x + y$
2. $y = -\frac{a}{b}x + \frac{P}{b}$
3. $-\frac{a}{b}$; a and b are constants, so $-\frac{a}{b}$ is constant.
4. The graphs have the same slope so they are parallel.
5. The y-intercepts, $\frac{P}{b}$, vary according to all possible values of P.
6. As you vary the y-intercept, the lines in the family $y = -\frac{a}{b}x + \frac{P}{b}$ will first, or last, touch the feasible region at a corner, or vertex. Thus, of all the lines containing a feasible point, the ones containing vertices will give the greatest and least values for $\frac{P}{b}$, and thus the greatest and least (feasible) values for P.
7. Yes.
8. Yes for both plots.

Name ______________________ Class ______________ Date ______________

Linear Programming

Activity 53

FILES NEEDED: Inequality Graphing App, Transformation Graphing App
Programs: A2L34B, A2L34C

In this activity you use the Inequality Graphing App to simplify the solving of linear programming problems.

Run the Inequality Graphing App. Run A2L34B and set up the Y= screen as shown below left. Use the SHADES menu (press ALPHA F2) to display a "clean" feasibility region, below right, and help you maximize the objective function $Profit = 0.75x + y$.

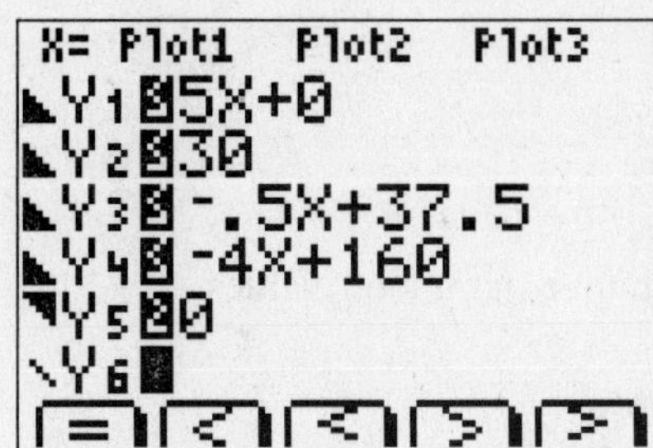

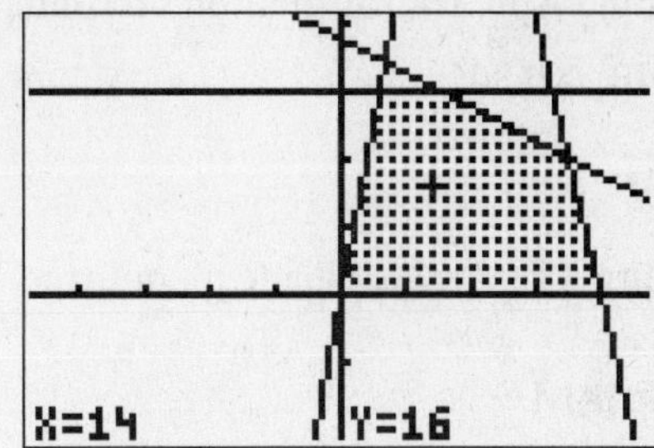

1. Use Pol-Trace to find coordinates of the five corner points. Use the cursor to find coordinates of three other feasible points. Record the coordinates in the table and complete the table.

Coordinates of point					(14, 16)		
Value of objective function: *Profit* $= 0.75x + y$					\$26.50		

2. Which point gives the greatest profit? The least profit? How much in each case?

3. Run the Transformation Graphing App and run A2L34C. The startup screen plots the feasibility corner points and a graph of the objective function $P = -0.75x + B$. Translate the graph vertically so it touches some part of the feasibility region but its y-intercept is as large as possible. What value of B gives the greatest y-intercept? At what feasibility point?

4. How do your answers to Questions 2 and 3 compare?

5. Which feasibility point gives maximum profit and what is the maximum profit for a new objective function $P = 2x + y$ or $y = -2x + P$? For a third objective function $P = 10x + y$? (*Hint:* Change the value of A. Then move the graph.)

6. You have used two methods to find the maximum profit. Using the table, you evaluated the objective function at each feasible corner point. Describe the other method.

7. Use the Inequality Graphing App to solve the following linear programming problem. Constraints: $y - x \leq 20$, $3y + x \leq 100$, $2y - x \geq 0$, $y + 0.5x \geq 20$; Objective function to maximize and minimize: $P = y - 2x$.

Linear Programming

Teacher Notes

Activity Objective

Students use the Inequality Graphing and Transformation Graphing Apps to solve linear programming problems.

Time

- 40–45 minutes

Materials/Software

- Inequality Graphing App, Transformation Graphing App
- Programs: A2L34B, A2L34C
- Activity worksheet

Skills Needed

- graph an inequality
- shade an intersection
- change parameter values

Classroom Management

- Students can work individually or in pairs depending on the number of calculators available.

Notes

- Press GRAPH to restore Shades and PoI-Trace at the bottom of a graph screen.
- PoI in PoI-Trace stands for "points of interest."

Answers

1. Non-vertex coordinates may vary. Samples are given.

coordinates	(0, 0)	(40, 0)	(35, 20)	(15, 30)	(6, 30)	(14, 16)	(10, 20)	(20, 10)	(30, 18)
P value	$0	$30	$46.25	$41.25	$34.50	$26.50	$27.50	$25	$40.50

2. (35, 20) for P = $46.25; (0, 0) for P = $0
3. 46.25; (35, 20)
4. The greatest profit data in Question 2 match the answers to Question 3.
5. (35, 20), $90; (40, 0), $400
6. Answers may vary. Sample: You graph members of the family $y = ax + P$ and find the line that contains a feasibility point and has the greatest y-intercept.
7. Maximum: $P = 20$ at (0, 20); Minimum: $P = -60$ at (40, 20).

Name ______________________ Class ______________ Date ______________

Quadratic Translations I

Activity 54

FILES NEEDED: Transformation Graphing App
Program: A2L52

A2L52 graphs the quadratic function $y = ax^2 + bx + c$ as $Y1 = AX^2 + BX + C$. In this activity you will study the relationship between the vertex of the graph and the values of *a*, *b*, and *c*.

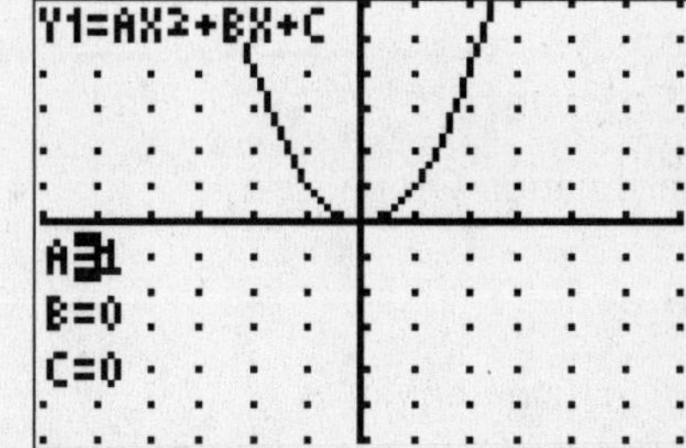

1. Run A2L52. The startup graph has A = 1, B = 0, and C = 0. Write the equation for this *parent* quadratic function. Find its vertex coordinates.

2. Do each of the following. Then describe the effects on the vertex and the *y*-intercept.

   a. Vary the value of A.

   b. Set A = 1 and vary the value of C.

   c. Set C = 2 and vary the value of B.

   d. Set A = 2 and B = 0. Then vary the value of C.

   e. Set C = −2 and vary the value of B.

   f. Set A = −4 and vary the value of B.

3. Generalize. Describe the effect on the graph of $y = ax^2 + bx + c$:

   a. when *a* changes from $a > 0$ to $a < 0$.

   b. when *c* changes in increments of 1 or −1.

   c. when $a = 1$ and *b* changes in increments of 1 or −1.

   d. when $a = 2$ and *b* changes in increments of 1 or −1.

4. The *x*-coordinate of the vertex is $-\frac{b}{2a}$. Set C = 0. Then choose three pairs of values for A and B and check the accuracy of this statement.

5. For each quadratic function below, state whether the parabola opens up or down. Also, find the vertex and *y*-intercept. Enter values for A, B, and C directly to check your work.

   a. $y = 2x^2 + 4x - 2$ b. $y = -3x^2 + 6x - 4$ c. $y = 2x^2 - 5x + 1$

Extension

6. What must be true for *a*, *b*, and/or *c* for a quadratic function to have a maximum value?

7. How could you find the maximum value? Use your method on 5b.

Quadratic Translations I

Activity Objective

Students use the Transformation Graphing App to explore how changing parameter values in the standard form, $y = ax^2 + bx + c$, of the quadratic function affects the graph of the function.

Time

- 15–20 minutes

Materials/Software

- Transformation Graphing App
- Program: A2L52
- Activity worksheet

Skills Needed

- change parameter values

Notes

- Students should uninstall the Transformation Graphing App when they complete the activity and then run DEFAULT.

Answers

1. $y = x^2$; $(0, 0)$

2. a. no effect
   b, d. Vertex and y-intercept are the same and move vertically by an amount equal to the change in C.
   c, e. Vertex traces the path of a parabola opening downward; no effect on y-intercept.
   f. Vertex traces the path of a parabola opening upward; no effect on y-intercept.

3. a. The parabola opens in the opposite direction.
   b. Graph moves vertically by an amount equal to the change in C.
   c. Each point of the graph traces the path of a parabola opening downward. Graph moves horizontally by a half unit for each unit change in B.
   d. Like part c, but now graph moves horizontally by a quarter unit for each unit change in B.

4. Check students' work.

5. a. up; $(-1, -4)$; -2 b. down; $(1, -1)$; -4 c. up; $\left(\frac{5}{4}, -2\frac{1}{8}\right)$; 1

6. a must be negative.

7. Let $x = -\frac{b}{2a}$. Substitute for x and solve for y.

Name ______________________ Class ______________ Date ______________

Quadratic Translations II

Activity 55

FILES NEEDED: Transformation Graphing App
Program: A2L53A

In "Quadratic Translations I" you studied translations of the quadratic function in standard form, $y = ax^2 + bx + c$. In this activity, you again study translations of the quadratic function. The translations will build from the parent function, $y = x^2$, to the *vertex form*, $y = a(x - h)^2 + k$.

A2L53A graphs the vertex form as $Y1 = A(X - B)^2 + C$. In the startup window, $A = 1$, $B = 0$, and $C = 0$, and the graph is the parent, $y = x^2$.

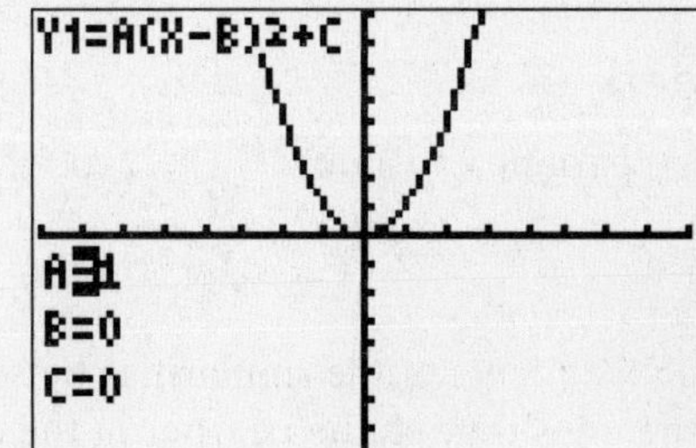

1. Run A2L53A. Vary the value of A. What is the effect on the graph of $Y1 = AX^2$ for increasing values of A? For decreasing values of A? What is the effect on the vertex for changing values of A?

2. What is the effect on the graph when you change values for A from $A > 0$ to $A < 0$?

3. Set $A = 1$. Vary the value of B. What is the effect on the graph of $Y1 = (X - B)^2$ for increasing values of B? For decreasing values of B? What is the effect on the vertex for changing values of B?

4. Predict the vertex x-coordinate for $B = 3$. For $B = -1$. Test your predictions.

5. With $B \neq 0$, vary the value of A. Compare the effects on the graph of $Y1 = A(X - B)^2$ with the effects on the graph of $Y1 = AX^2$ in Questions 1 and 2.

6. Vary the value of C. What is the effect on the graph of $Y1 = A(X - B)^2 + C$ for increasing values of C? For decreasing values of C? What is the effect on the vertex for changing values of C?

7. Copy and complete the following paragraphs to summarize what the vertex form of a quadratic function tells you about the graph.

For $a > 0$ in $y = a(x - h)^2 + k$, the parabola opens _?_. The parabola opens down for _?_. A positive value of h shifts the graph of the parent function to the _?_. A value $h < 0$ shifts the graph to the _?_. The value of h suggests a _?_ (horizontal, vertical) translation of the parent function by _?_ units. The value of k suggests a _?_ translation of the parent function by _?_ units.

The coordinates of the vertex of the graph of $y = a(x - h)^2 + k$ are (_?_, _?_).

Extension

8. Write each function in vertex form and describe the graph.

a. $y - 5 = -2(x - 3)^2$ **b.** $y = x^2 - 6x + 9$ **c.** $y = 2x^2 + 6x + 1$

Quadratic Translations II

Teacher Notes

Activity Objective

Students use the Transformation Graphing App to explore how changing parameter values in the vertex form, $y = a(x - h)^2 + k$, of the quadratic function affects the graph of the function.

Time

- 15–20 minutes

Materials/Software

- Transformation Graphing App
- Program: A2L53A
- Activity worksheet

Skills Needed

- change parameter values

Notes

- Remind students of the similarities between these transformations and the ones they studied earlier in the course.
- Compare (X + 3) with (X – B) and show that B must be −3.
- Students should uninstall the Transformation Graphing App when they complete the activity and then run DEFAULT.

Answers

1. Increasing A narrows the graph vertically. Decreasing A, A > 0, widens the graph. Changing values of A has no effect on the vertex.

2. If A > 0, the graph opens up. If A < 0, the graph opens down.

3. Increasing B moves the graph (and vertex) to the right. Decreasing B moves the graph (and vertex) to the left. The coordinates of the vertex are (B, 0).

4. 3; −1

5. The effects are the same.

6. Increasing C moves the graph (and vertex) up. Decreasing C moves the graph (and vertex) down.

7. up; A < 0; right; left; horizontal; $|h|$; vertical; $|k|$; h; k

8.
a. $y = -2(x - 3)^2 + 5$, parabola, vertex (3, 5), opens down
b. $y = (x - 3)^2$, parabola, vertex (3, 0), opens up
c. $y = 2(x + 1.5)^2 - 3.5$, parabola, vertex (−1.5, −3.5), opens up.

Name ______________________ Class ______________ Date ______________

Dodge 'Em

Activity 56

FILES NEEDED: Transformation Graphing App
Program: A2L53B

You are the captain of a space freighter. You have to make deliveries to space stations A, B, and C.

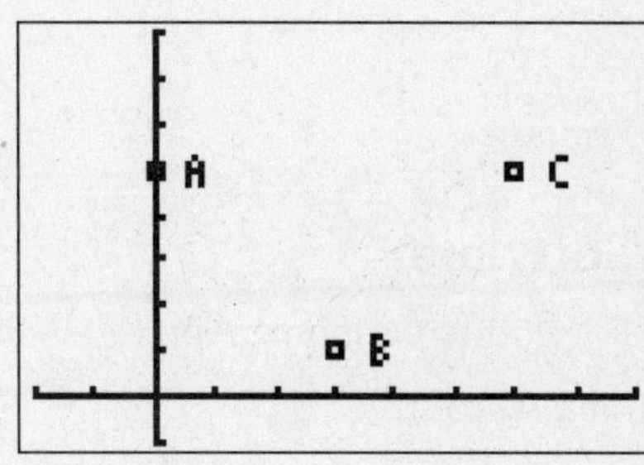

Unfortunately, space is not clear. There are asteroids (+) in the plane of your path. You must not collide with an asteroid.

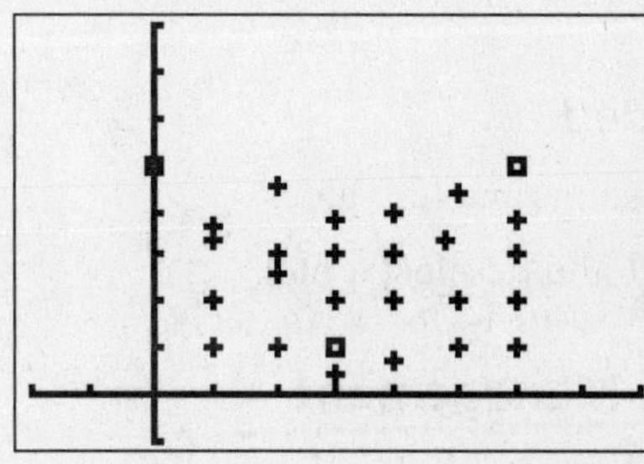

There is a safe parabolic path that runs through the asteroid field and connects the space stations.

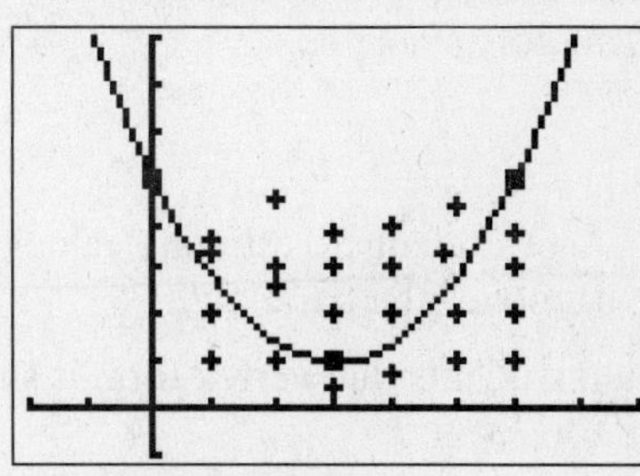

You must find it.

A2L53B graphs the vertex form $y = a(x - h)^2 + k$ of a quadratic function as $\text{Y1} = \text{A}(\text{X} - \text{B})^2 + \text{C}$ with A = 1, B = 0, and C = 0.

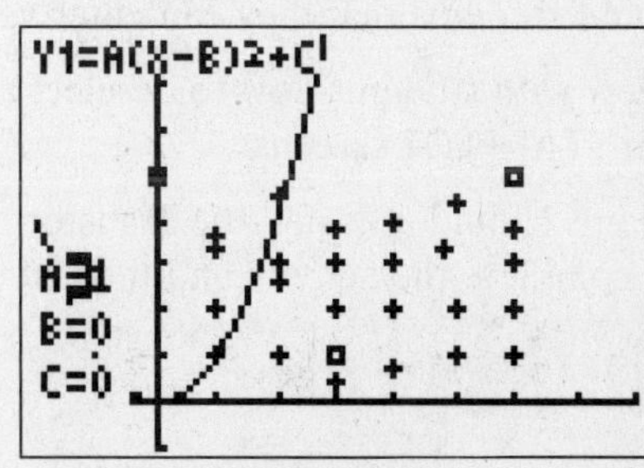

Run A2L53B and find values of A, B, and C that will miss the asteroids. Record the vertex form of the successful quadratic function in your Captain's Log.

Extension

Switch from Plot1 to Plot2 (keeping Plot3 active) and press GRAPH for a more challenging space flight. Find a function to guide your spacecraft safely to the three space stations.

Dodge 'Em

Activity Objective

Students use the Transformation Graphing App to practice translating the graph of a parabola by changing the parameter values in Y1 = A(X − B)^2 + C.

Time

- 10–15 minutes

Materials/Software

- Transformation Graphing App
- Program: A2L53B
- Activity worksheet

Skills Needed

- change parameter values
- select and deselect a plot

Classroom Management

- Students can work individually or in pairs depending on the number of calculators available.

Notes

- Encourage students to estimate values for A, B, and C before they start to "move" the parabola.
- Remind students that vertex form is shown in the text as $y = a(x - h)^2 + k$
- Remind students that they can enter values for A and B directly.
- Review with students how to deselect and select a plot in either the Y= or STAT PLOT screens.
- Students should uninstall the Transformation Graphing App when they complete the activity and then run DEFAULT.

Answers

Answers may vary. Samples: $y = 0.45(x - 3) + 1$ for Plot1; $y = 0.5(x - 3.3) + 1.4$ for Plot2.

Name ______________________ Class ______________ Date ______________

Quadratic Function Match II

Activity 57

FILES NEEDED: Transformation Graphing App
Program: A2L53C

A2L53C graphs the vertex form of the quadratic function, $y = a(x - h)^2 + k$, as $Y1 = A(X - B)^2 + C$ with $A = 1$, $B = 0$, and $C = 0$.

In this activity you see three plots, one at a time. You are to change the values of A, B, and C to find a perfect match for the given plot.

1. Run A2L53C. Find a quadratic function whose graph matches the plot.

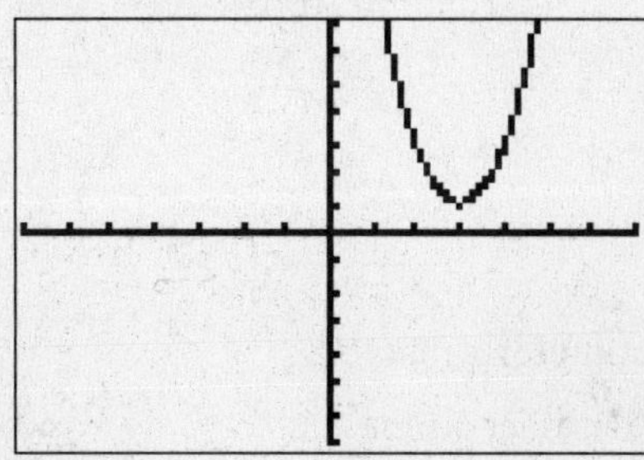

2. Switch from Plot1 to Plot2. Press GRAPH to see A2L53C for the second plot. Find a matching quadratic function.

```
Plot1 Plot2 Plot3
\Y1=A(X-B)²+C
\Y2=
\Y3=
\Y4=
\Y5=
\Y6=
\Y7=
```

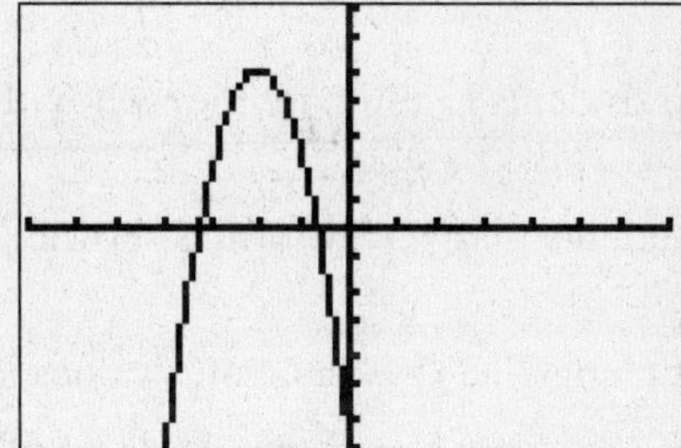

3. Switch from Plot2 to Plot3. Press GRAPH to see A2L53C for the third plot. Find a matching quadratic function.

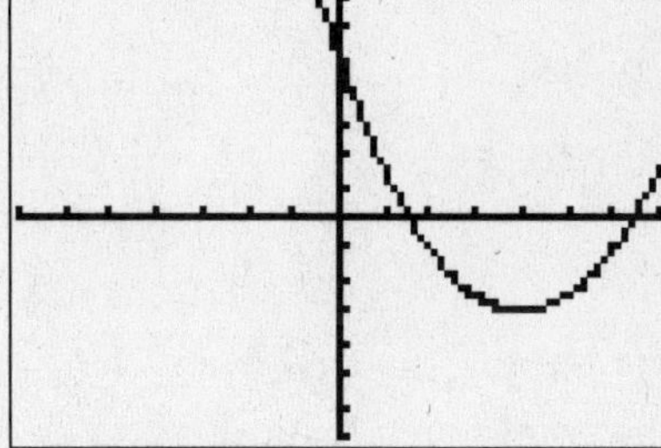

4. Write the three functions found in Questions 1–3 in standard form. Draw their graphs in the respective Plots of A2L53C. How can you tell whether your standard form is correct?

Quadratic Function Match II

Activity Objective

Students use the Transformation Graphing App to practice translating the graph of a parabola by changing the parameter values in $Y1 = A(X - B)^2 + C$.

Time

- 10–15 minutes

Materials/Software

- Transformation Graphing App
- Program: A2L53C
- Activity worksheet

Skills Needed

- change parameter values
- select and deselect a plot

Classroom Management

- Students can work individually or in pairs depending on the number of calculators available.

Notes

- Encourage students to estimate the values of A, B, and C before they start to "move" the parabola.
- Remind students that vertex form is shown in the text as $y = a(x - h)^2 + k$.
- Remind students that they can enter values for A and B directly.
- Review with students how to deselect and select a plot in either the Y= or STAT PLOT screens.

Answers

1. $y = 2(x - 3)^2 + 1$
2. $y = -3(x + 2)^2 + 5$
3. $y = 0.5(x - 4)^2 - 3$
4. $y = 2x^2 - 12x + 19$; $y = -3x^2 - 12x - 7$; $y = 0.5x^2 - 4x + 5$; It is correct if the graphs are identical.

Name ______________________ Class ______________ Date ______________

Follow the Bouncing Ball I

Activity 58

FILES NEEDED: Transformation Graphing App
Program A2L53D

For this activity, a motion detector collected data about a bouncing ball. The graph at the right shows a plot of the data. You are to find models for some of the data.

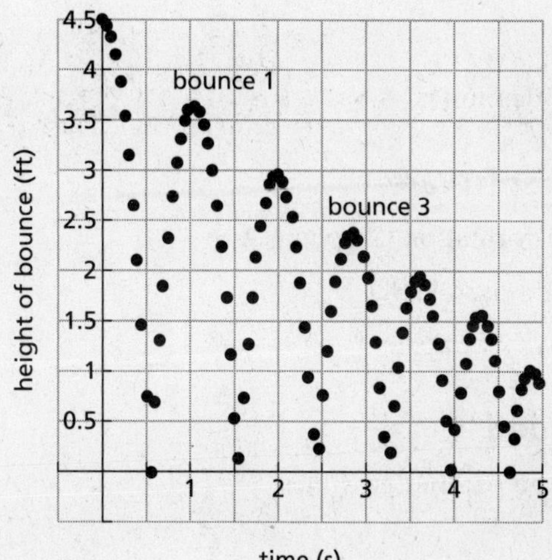

1. Use the graph. On which bounce did the ball go the highest? How do you know?
2. About how high did the ball go on its first bounce?
3. Approximately how long was the ball in the air on its first bounce? On its fifth bounce?
4. Run A2L53D. It graphs the vertex form of the quadratic function, $y = a(x - h)^2 + k$, as $Y1 = A(X - B)^2 + C$. Change the values of A, B, and C to find a model for the first bounce. Write the function equation for your model.
5. Find a function equation that models the third bounce.
6. For the first and third bounces, how are the two values of A related?
7. How do the two values of B compare? The two values of C?
8. What do you suppose are the "real world" meanings of B and C?

Extension

9. Do you think the motion detector found the actual vertex point on each bounce? Explain.
10. The pictures below show the data collected during the second bounce. Do these plots show the actual vertex? Explain.
11. If the motion detector does not find the actual vertex, how could you find it?

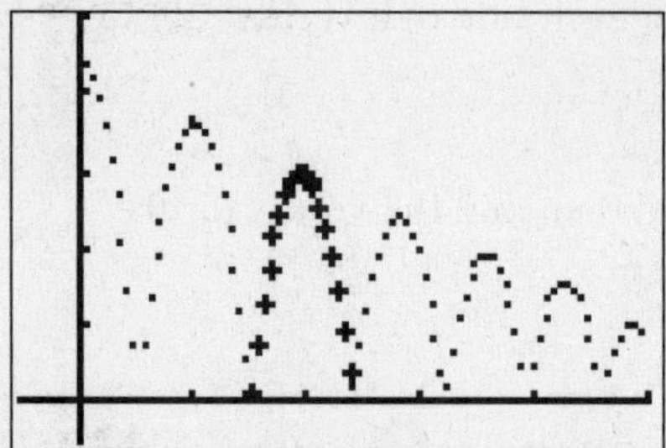

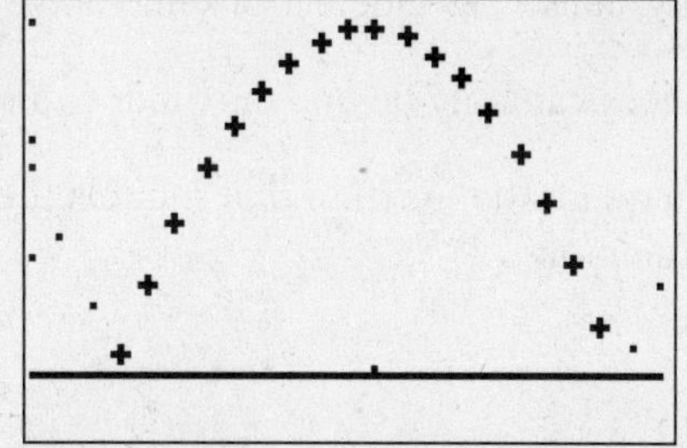

Follow the Bouncing Ball I

Activity Objective

Students use the Transformation Graphing App to model the bounce of a ball based on bounce data gathered by a motion detector.

Time

- 20–30 minutes

Materials/Software

- Transformation Graphing App
- Program: A2L53D
- Activity worksheet

Skills Needed

- change parameter values

Classroom Management

- Students can work individually or in pairs depending on the number of calculators available.

Notes

- Students should realize while answering Question 6 that the A values should be the same if they aren't already.

Answers

1. bounce 1; it has the highest peak

2. 3.7 ft

3. 1 s; 0.5 s

4–6. Answers may vary. Samples are given.

4. $y = -14.9(x - 1.05)^2 + 3.7$

5. $y = -14.9(x - 2.84)^2 + 2.5$

6. They are the same.

7. B is greater and C is less on the third bounce.

8. B is the horizontal distance from the starting position. C is the height of the bounce.

9. Not necessarily; consecutive detections may have occurred on each side of a vertex.

10. No, because there can only be one maximum value.

11. You can find a quadratic function that models the bounce and then find the vertex for the graph of the function.

Name ____________ Class ____________ Date ____________

Graphs, Zeros, and Factors

Activity 59

FILES NEEDED: Transformation Graphing App
Program: A2L62A

In A2L62A, the polynomial function $y = (x - a)(x - b)(x - c)$ is in *factored form* and is graphed as Y1 = (X − A)(X − B)(X − C).

In this activity you will explore relationships among

- the x-intercepts or *zeros* of a polynomial function,
- the solutions of the related polynomial equation, and
- the factors of the related polynomial expression.

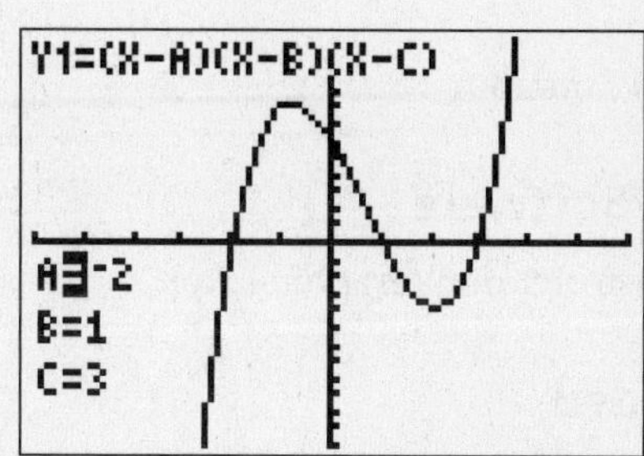

1. Run A2L62A. What are the values of A, B, and C in the startup window? What is the function whose graph is shown?

2. The polynomial function $y = (x - a)(x - b)(x - c)$ has $(x - a)(x - b)(x - c) = 0$ as its *related polynomial equation.* What is the related polynomial equation for the function you found in Question 1?

3. Explain both algebraically and graphically why a, b, and c are the zeros of the polynomial function $y = (x - a)(x - b)(x - c)$. What are the zeros of the polynomial function you found in Question 1?

4. $y = ax^3 + bx^2 + cx + d$ is the *standard form* of the function $y = (x - a)(x - b)(x - c)$. (*Note:* a, b, and c in the two forms do not represent the same values.) Write the standard form of the polynomial function you found in Question 1. Write its corresponding polynomial equation. What are the solutions of this equation? Explain how you know.

5. Change the value of A as indicated. Describe the effect on the graph and on the solutions of each related polynomial equation.

a. Set A = −1 **b.** Set A = 0 **c.** Set A = 1

6. Draw a sketch to predict what the graph of each function will look like. Use A2L62A to test your predictions.

a. $y = (x - 1)(x - 2)(x - 3)$ **b.** $y = (x - 1)^2(x - 3)$ **c.** $y = (x - 1)^3$

7. For each polynomial, graph the function. Use the graph to factor the polynomial. Solve the related polynomial equation.

a. $x^3 - 2x^2 - 16x + 32$ **b.** $x^3 + 8x^2 - 4x - 32$ **c.** $x^4 - 3x^3 - 15x^2 + 19x + 30$

d. $x^4 - 9x^2 + 4x + 12$ **e.** $-x^3 - 2x^2 + 5x + 6$ **f.** $x^4 - 5x^3 + 6x^2 + 4x - 8$

Extension

8. How many solutions are possible for a polynomial equation that uses a polynomial of degree 2? Degree 3? Degree 4? Degree n?

Graphs, Zeros, and Factors

Teacher Notes

Activity Objective

Students use the Transformation Graphing App to explore the relationships among the x-intercepts of a polynomial function, the zeros of the function, the solutions of the related polynomial equation, and the factors of the related polynomial expression.

Time

- 40–45 minutes

Materials/Software

- Transformation Graphing App
- Program: A2L62A
- Activity worksheet

Skills Needed

- change parameter values

Notes

- Remind students that they can enter A and B values directly.
- Students should uninstall the Transformation Graphing App when they complete the activity and then run DEFAULT.

Answers

1. $-2, 1, 3$; $y = (x + 2)(x - 1)(x - 3)$

2. $(x + 2)(x - 1)(x - 3) = 0$

3. a, b, and c are zeros because $y = 0$ when $x = a$, $x = b$, or $x = c$. Graphically, a, b, and c are the x-intercepts (where $y = 0$). $-2, 1, 3$

4. $y = x^3 - 2x^2 - 5x + 6$; $x^3 - 2x^2 - 5x + 6 = 0$; This equation is equivalent to the Question 2 equation, so their solutions are the same, $-2, 1, 3$.

5.
- **a.** A zero and a solution change from -2 to -1.
- **b.** A zero and a solution change to 0.
- **c.** The graph touches but does not cross the x-axis at $x = 1$. There are only two solutions, 1 and 3.

6. Check students' graphs. **a.** zeros at 1, 2, 3 **b.** See Question 5, part c. **c.** The graph intersects the x-axis once—at $(1, 0)$.

7. **a.** $-4, 2, 4$ **b.** $-8, -2, 0$ **c.** $-3, -1, 2, 5$ **d.** $-3, -1, 2$ **e.** $-3, -1, 2$ **f.** $-1, 2$

8. $2; 3; 4; n$

Name ______________________ Class ______________ Date ______________

Polynomial Function Match

Activity 60

FILES NEEDED: Transformation Graphing App
Program: A2L62B

A2L62B graphs the factored form, $y = (x - a)(x - b)(x - c)$, of the polynomial function as Y1 = (X – A)(X – B)(X – C).

In this activity you see three plots, one at a time. You are to change the values of A, B, and C to find a perfect match for the given plot.

1. Run A2L62B. Find a polynomial function whose graph matches the plot.

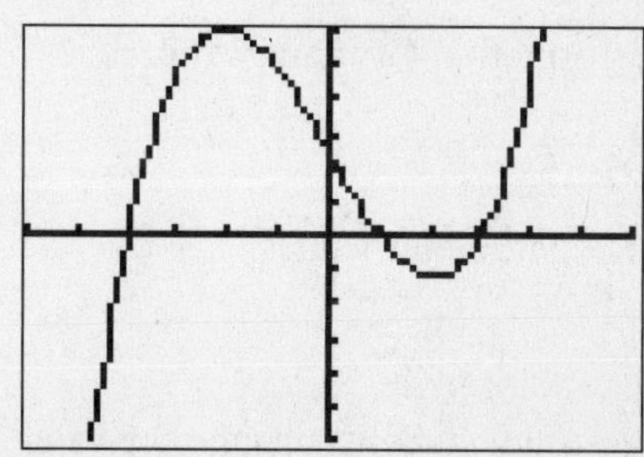

2. Switch from Plot1 to Plot2. Press GRAPH to see A2L62B for the second plot. Find a matching polynomial function.

```
 Plot1 Plot2 Plot3
\Y1=(X-A)(X-B)(X
-C)
\Y2=
\Y3=
\Y4=
\Y5=
\Y6=
```

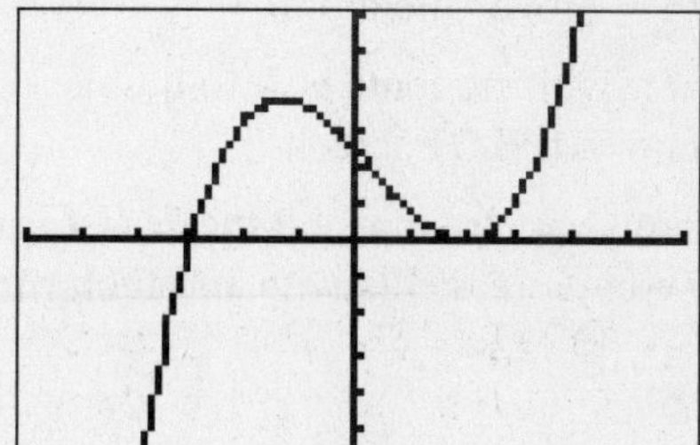

3. Switch from Plot2 to Plot3. Press GRAPH to see A2L62B for the third plot. Find a matching polynomial function.

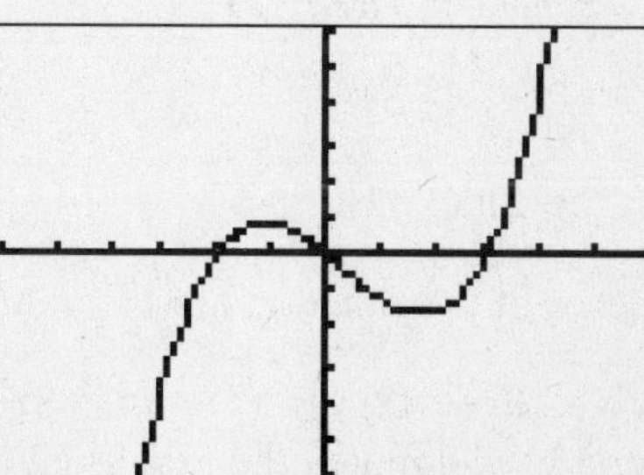

4. Explain how you could have answered Questions 1–3 without manipulating values for A, B, and C.

5. Write the three functions found in Questions 1–3 in standard form. Draw their graphs in the respective Plots of A2L62B. How can you tell whether your standard form is correct?

Extension

6. For all three graphs above, explain the behavior of the parts of the graphs that are not shown (sometimes called *end behavior*). What could you do to one of the given functions to make this type of behavior occur in Quadrants II and IV instead of I and III?

Polynomial Function Match

Teacher Notes

Activity Objective

Students use the Transformation Graphing App to explore how changing parameter values in the factored form $y = (x - a)(x - b)(x - c)$ of the polynomial function affects the graph of the function.

Time

- 25–30 minutes

Materials/Software

- Transformation Graphing App
- Program: A2L62B
- Activity worksheet

Skills Needed

- change parameter values
- deselect and select a plot

Notes

- Suggest that students change the value of A to find which graph is the "movable" one.
- Encourage students to estimate the values of A, B, and C before they start to "move" the graph.
- Review with students how to deselect and select a plot in either the Y= or STAT PLOT screens.
- Students should uninstall the Transformation Graphing App when they complete the activity and then run DEFAULT.

Answers

1. $y = (x + 4)(x - 1)(x - 3)$
2. $y = (x + 3)(x - 2)^2$
3. $y = x(x + 2)(x - 3)$
4. The zeros of the function are the values a, b, and c.
5. $y = x^3 - 13x + 12; y = x^3 - x^2 - 8x + 12; y = x^3 - x^2 - 6x;$ The graphs will match the graphs given in the Plots.
6. As $|x|$ becomes arbitrarily large, $|y|$ becomes arbitrarily large. You can reflect the graph across the x-axis by introducing a factor of -1.

Name ______________________ Class ______________ Date ______________

Radical Translations I

Activity 61

FILES NEEDED: Transformation Graphing App
Program: A2L78A

In the "Quadratic Translations II" Activity, you studied translations of the quadratic function, building from the parent function $y = x^2$ to a general form that you can write as $y = a(x - b)^2 + c$.

In this activity you will study translations of the radical function, building from the parent function $y = \sqrt{x}$ to the general form $y = a\sqrt{x - b} + c$. You should notice that the two general forms are quite similar. It turns out that the effects of a, b, and c on the graphs of the functions are quite similar as well.

A2L78A graphs the square root function $y = a\sqrt{x - b} + c$ as Y1= A√(X − B) + C.

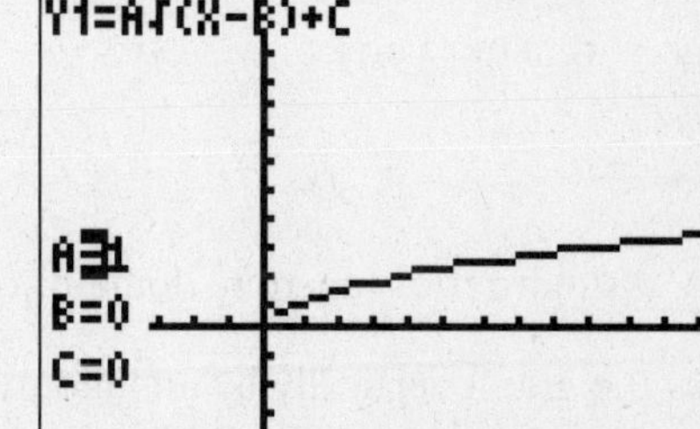

1. Run A2L78A. Write the function equation for the graph shown in the startup window. What special name does this function have? Describe its domain and range.
2. Recall the effects of change in c on the graph of $y = a(x - b)^2 + c$. Predict how change in c will affect the graph of $y = a\sqrt{x - b} + c$.
3. Test your prediction. Change the value of C in increments of 1. What happens to the graph? Use increments of −1. What happens to the graph?
4. Recall the effect of change in b on the graph of $y = a(x - b)^2 + c$. Predict how change in b will affect the graph of $y = a\sqrt{x - b} + c$.
5. Test your prediction. Set C = 0. Change the value of B in increments of 1. What happens? In increments of −1. What happens?
6. For each given function, what are the values of c and b? Describe the translation of the parent graph. Give the domain and range.

   a. $y = \sqrt{x} + 2$ **b.** $y = \sqrt{x} - 3$ **c.** $y = \sqrt{x - 3}$ **d.** $y = \sqrt{x + 2}$
7. For the function $y = \sqrt{x - 3} + 4$, describe two translations of the parent graph that result in the graph of this function. Give the domain and range. Use A2L78A to check your work.
8. Predict how changes in a will affect the graph of $y = a\sqrt{x - b} + c$. Use A2L78A to change A and check your predictions.
9. Shift the graph of $y = \sqrt{x}$ by 2 units down and 3 units left. What is the function equation for the new graph. Check using A2L78A.
10. Describe how the graph of each relates to the graph of $y = \sqrt{x}$.

    a. $y = 2\sqrt{x - 5} + 4$ **b.** $y = \frac{1}{2}\sqrt{x + 1} - 4$ **c.** $y = -4\sqrt{x + 3} + 6$

Radical Translations I

Teacher Notes

Activity Objective

Students use the Transformation Graphing App to explore how changing parameter values in the square root function $y = a\sqrt{x - b} + c$ affects the graph of the function.

Time

- 40–45 minutes

Materials/Software

- Transformation Graphing App
- Program: A2L78A
- Activity worksheet

Skills Needed

- change parameter values

Answers

1. $y = \sqrt{x}$; square root function; domain: $x \geq 0$, range: $y \geq 0$.
2. Moves the graph vertically by an amount equal to the change in c.
3. Moves up in steps of 1; moves down in steps of 1.
4. Moves the graph horizontally by an amount equal to the change in b.
5. Moves right in steps of 1; moves left in steps of 1.
6. a. 2, 0; moves up 2 units; domain: $x \geq 0$, range: $y \geq 2$.

   b. -3, 0; moves down 3 units; domain: $x \geq 0$, range: $y \geq -3$.

   c. 0, 3; moves right 3 units; domain: $x \geq 3$, range: $y \geq 0$.

   d. 0, -2; moves left 2 units; domain: $x \geq -2$, range: $y \geq 0$.
7. Moves up 4 units and right 3 units; domain: $x \geq 3$, range: $y \geq 4$.
8. Increase in a will stretch the graph vertically. Decrease in a ($a > 0$) will shrink the graph vertically. A sign change in a will reflect the graph across the line $y = c$.
9. $y = \sqrt{x + 3} - 2$
10. For each of a–c, begin with the graph of $y = \sqrt{x}$.
    a. Shift right 5, stretch vertically by factor of 2, shift up 4.
    b. Shift left 1, shrink vertically by factor of $\frac{1}{2}$, shift down 4.
    c. Shift left 3, stretch vertically by factor of 4, reflect across x-axis, shift up 6.

Name ______________________ Class ______________ Date ______________

Radical Translations II

Activity 62

> **FILES NEEDED:** Transformation Graphing App
> Program: A2L78B

In "Radical Translations I" you studied translations of the square root function, building from the parent function $y = \sqrt{x}$ to the general form $y = a\sqrt{x - b} + c$. You can take the general form further—to the *n*th-root function, $y = a\sqrt[n]{x - b} + c$.

```
Plot1 Plot2 Plot3
\Y1 =D ×√(X)
\Y2 =A ³√(X−B)+C
\Y3 =A4 ×√(X−B)+C
\Y4 =
\Y5 =
\Y6 =
\Y7 =
```

A2L78B graphs the *n*th-root function using Y1, Y2, and Y3, as shown at the right.

1. For each of Y1, Y2, and Y3, complete the following using *a*, *b*, *c*, and *n*.

$y_1 = \underline{\ ?\ }$ $\quad y_2 = \underline{\ ?\ }$ $\quad y_3 = \underline{\ ?\ }$

2. Run A2L78B. The startup screen shows the parent function $y = \sqrt{x}$. (Why?) Change the index D (or *n*). Write equations for the parent functions you see for D = 3, 4, 5, . . . Describe any pattern you see in the sequence of displayed graphs.

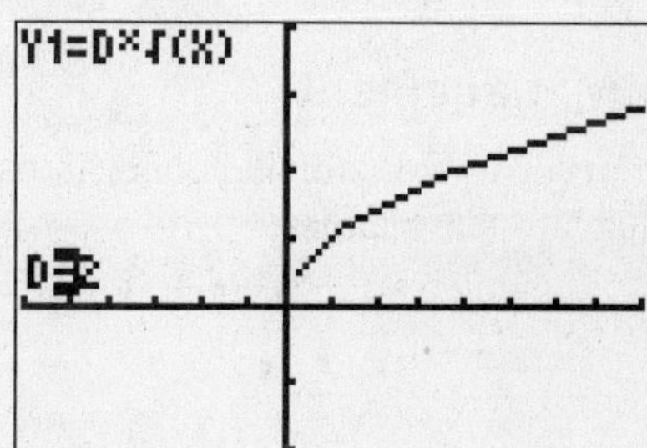

3. Graphs of quadratic and absolute value functions have a *vertex* that you can use as a central point for describing translations. What point can you use as the *central translation point* for a parent radical function whose index is an odd number? An even number?

4. What are the domain and range of the parent *n*th-root function $y = \sqrt[n]{x}$ when *n* is even? When *n* is odd?

5. As you might suspect, the effects of *a*, *b*, and *c* on the graphs of $y = a\sqrt{x - b} + c$ and $y = a\sqrt[n]{x - b} + c$ are the same. Use what you know about the square root function to predict how the graph of the parent cube root function translates to the graph of each function given below.

a. $y = \sqrt[3]{x - 3} + 1$ **b.** $y = \sqrt[3]{x + 1} - 2$ **c.** $y = -\sqrt[3]{x} + 2$

Use A2L78B to test your predictions. In the Y= window (see top of page), select Y2 by highlighting its "=" sign. Then press GRAPH.

6. For each function below, list the domain and range, and sketch a graph. Use Y2 and Y3 in A2L78B to check your sketches.

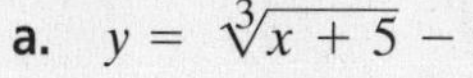

a. $y = \sqrt[3]{x + 5} - 2$ **b.** $y = \sqrt[4]{x} + 3$

c. $y = -\sqrt[3]{x + 2} + 3$ **d.** $y = 2\sqrt[4]{x + 3}$

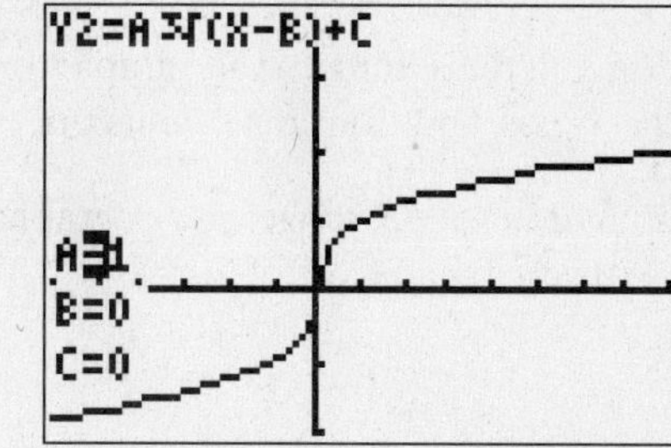

Radical Translations II

Teacher Notes

Activity Objective

Students use the Transformation Graphing App to study the effect of n on the graphs of the parent radical function $y = \sqrt[n]{x}$, and to explore how changing other parameter values in $y = a\sqrt[n]{x - b} + c$ affects the graph of the function.

Time

- 40–45 minutes

Materials/Software

- Transformation Graphing App
- Program: A2L78B
- Activity worksheet

Skills Needed

- change parameter values

Classroom Management

- Students can work individually or in pairs depending on the number of calculators available.

Answers

1. $y_1 = \sqrt[n]{x}; y_2 = a\sqrt[3]{x - b} + c; y_3 = a\sqrt[4]{x - b} + c$

2. $y = \sqrt[3]{x}, y = \sqrt[4]{x}, y = \sqrt[5]{x}, \ldots$; Answers may vary. Sample: When n is odd, the graphs of $y = \sqrt[n]{x}$ show similar behavior in Quadrants I and III. When n is even, the graphs show similar behavior in Quadrant I and don't appear anywhere else.

3. $(0, 0); (0, 0)$

4. nonnegative real numbers; real numbers

5. **a.** Translates right 3 and up 1.

b. Translates left 1 and down 2.

c. Reflects across the x-axis and translates up 2.

6. **a.** domain: reals; range: reals; graph: translate $y = \sqrt[3]{x}$ by 5 units left and 2 units down.

b. domain: $x \geq 0$; range: $y \geq 3$; graph: translate $y = \sqrt[4]{x}$ by 3 units up.

c. domain: reals; range: reals; graph: translate $y = \sqrt[3]{x}$ by 2 units left, reflect across the x-axis, and translate 3 units up.

d. domain: $x \geq -3$; range: $y \geq 0$; graph: translate $y = \sqrt[4]{x}$ by 3 units left and stretch vertically by a factor of 2.

Name ____________________ Class ____________ Date ____________

Choose a Model II

Activity 63

FILES NEEDED: Transformation Graphing App
Program: A2L81A

In A2L81A, you will view three plots, one at a time. You are to select the appropriate family of functions and then use the Transformation Graphing App to find a family member that models the displayed data.

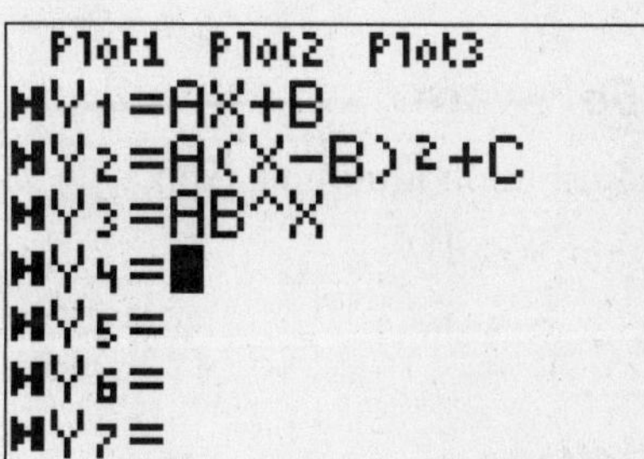

Function Families

Linear	Quadratic	Exponential
$y = mx + b$	$y = a(x - b)^2 + c$	$y = ab^x$
Y1 = AX + B	Y2 = A(X − B)^2 + C	Y3 = AB^X

The numbers *a*, *b*, and *c* are called *parameters*. For each plot there is a "perfect" model. Choose Y1, Y2, or Y3 and be prepared to take parameter values to two decimal places to find the perfect model.

1. Run A2L81A. In the Y= window, select the function you think will give the best model by highlighting its "=" sign. Then press GRAPH. Change the parameter values until you have the graph passing through the plotted points. Write the function for your graph.

2. Switch from Plot1 to Plot2. Press GRAPH to see the second plot. Select a function. Change its parameter values to get the perfect model. Write the function for your graph.

Plot1 Plot2 Plot3
Y1=AX+B
Y2=A(X−B)²+C
Y3=AB^X
Y4=
Y5=
Y6=
Y7=

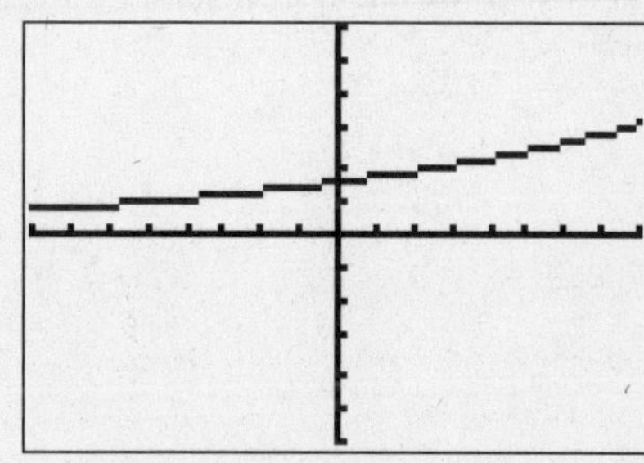

3. Switch from Plot2 to Plot3.

   Press GRAPH to see the third plot. Select a function. Change its parameter values to get the perfect model. Write the function for your graph.

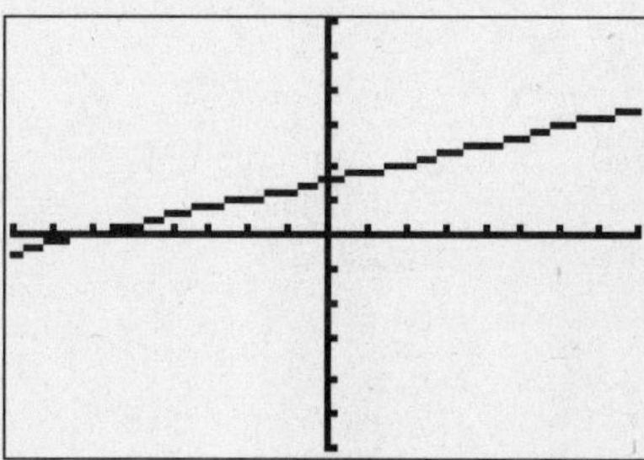

Extension

4. Suppose you had to make a "Choose a Model" challenge for a classmate. How would you construct the given plot?

Choose a Model II

Activity Objective

Students make a visual choice of an appropriate model and then use the Transformation Graphing App to confirm their choice.

Time

- 20–25 minutes

Materials/Software

- Transformation Graphing App
- Program: A2L81A
- Activity worksheet

Skills Needed

- change parameter values
- select and deselect a plot

Classroom Management

- Students can work individually or in pairs depending on the number of calculators available.

Notes

- Review with students how to deselect and select a plot in either the Y= or STAT PLOT screens.

Answers

1. $y = -2(x + 5)^2 + 3$

2. $y = 1.5(1.1)^x$

3. $y = 0.25x + 1.5$

4. Check students' work.

Name ______________________ Class ______________ Date ______________

Follow the Bouncing Ball II

Activity 64

FILES NEEDED: Transformation Graphing App
Program: A2L81B

The graph at the right shows the plot of the motion-detector data gathered for a bouncing ball and used in "Follow the Bouncing Ball I." In that activity you found equations that modeled some of the bounces.

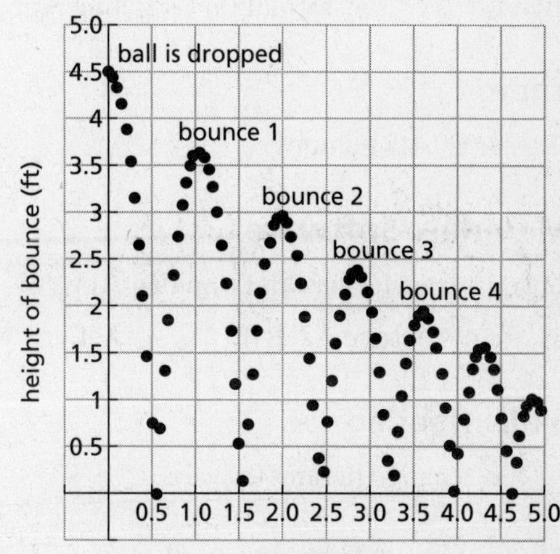

In this activity you will find a relationship between bounce number and height. Based on the graph:

1. From what height was the ball dropped?

2. How high was bounce 1?

3. How long was the ball in the air during bounce 1?

4. Complete the following table. Heights are in feet.

Bounce number	Maximum bounce height	Difference of this height from previous height	Ratio of this height to previous height
0	4.5		
1	3.7	$4.5 - 3.7 = 0.8$	$\frac{3.7}{4.5} \approx 0.82$
2			
3			
4			
5			

5. Study the data in the table. Of the functions you have studied so far—linear, absolute value, quadratic, radical, and exponential—which do you think would make the best model for the relationship between the bounce number and the bounce height? Explain. Then write a function equation that you feel would be a good model for the relationship between bounce number and bounce height.

6. Run A2L81B. Trace on the plot and compare with the first two columns of your table. If close, you do not need to change your table. Choose Y1, Y2, or Y3 (highlight the "=" sign and press ENTER) to match your choice of function in Question 5. Press GRAPH and change the A, B, C values to make the best model you can. Record the function equation for your model.

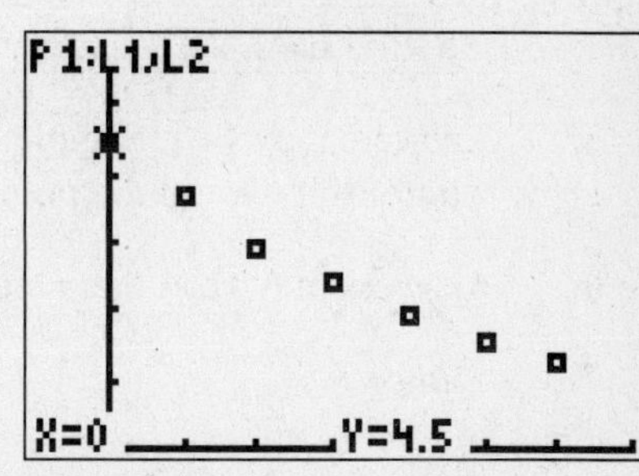

7. Use your model. When will the bounce height be less than one foot?

8. According to your model, will the ball ever stop bouncing? Explain.

Follow the Bouncing Ball II

Teacher Notes

Activity Objective

Students make a choice of an appropriate model based on tabular data and then use the Transformation Graphing App to confirm their choice.

Time

- 20–25 minutes

Materials/Software

- Transformation Graphing App
- Program: A2L81B
- Activity worksheet

Skills Needed

- change parameter values

Classroom Management

- Students can work individually or in pairs depending on the number of calculators available.

Notes

- Review with students how to deselect and select a plot in either the Y= or STAT PLOT screens.

Answers

1. 4.5 ft

2. about 3.7 ft

3. 1 s

4. Answers may vary. Sample:

2	2.9	0.8	0.78
3	2.4	0.5	0.83
4	1.9	0.5	0.79
5	1.6	0.3	0.84

5. Answers may vary. Sample: Exponential; Only for an expontential relationship would the points lie on a curve with an asymptote; $y = 4.5(0.8)^x$

6. Answers may vary. Sample: $y = 4.5(0.8)^x$.

7. bounce 7

8. No; $y > 0$ for all x.

Name ______ Class ______ Date ______

Asymptotes for Exponentials

Activity 65

FILES NEEDED: Transformation Graphing App
Programs: A2L82A, A2L82B

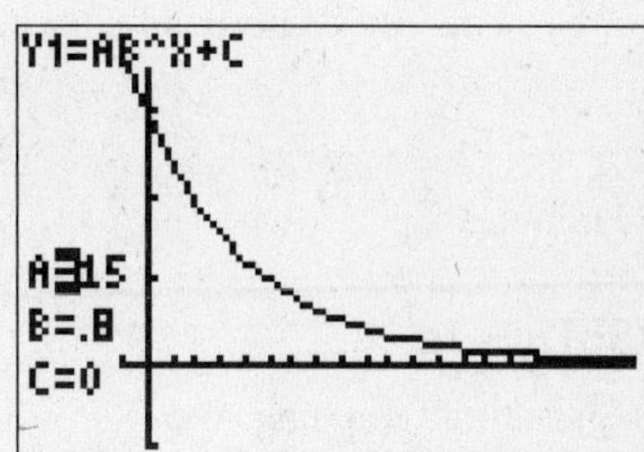

A2L82A graphs the exponential function $y = ab^x + c$ as Y1 = AB^X + C.

In this activity you will investigate the effects of changes in a, b, and c on the graph of the exponential function and its asymptote.

1. Run A2L82A. Write the function equation for the startup graph. Explain why the function is decreasing as x gets large. Which line is a horizontal asymptote for the graph? Explain.

2. Predict what will happen to the asymptote of the graph when you vary the value of C. Test your prediction. (*Note:* On the y-axis the scale marks are 5 units apart.)

3. When you vary C, you change the y-intercept as well as the asymptote. For $y = ab^x + c$ in general, what is the equation of the asymptote? What is the y-intercept? Use A2L82A and the following functions to test your answers.

   a. $y = 8(0.3)^x + 4$ **b.** $y = 10(0.9)^x + 2$ **c.** $y = 10(0.8)^x - 5$ **d.** $y = 15(0.75)^x - 4$

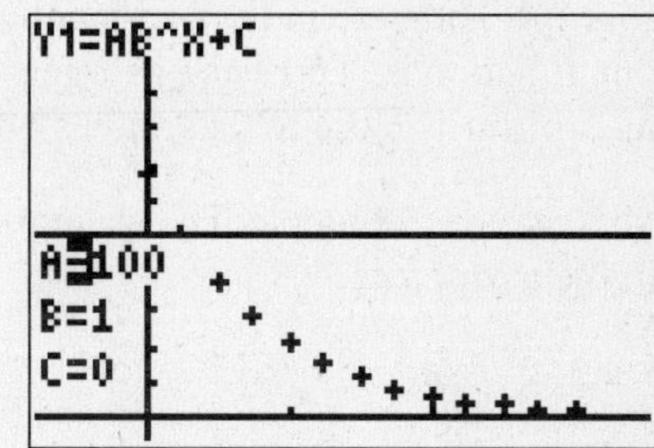

4. Run A2L82B. The startup screen shows a scatter plot for the "Cooling Coffee" data on page 430 of your textbook. Vary A, B, and C to find a function model for the data. (Use two decimal places for B.) Give the asymptote and the y-intercept. Tell what each means in terms of the cooling coffee.

5. If the coffee was 135°F above room temperature and room temperature was 72°F, what would be the coffee temperature?

6. Switch from Plot1 to Plot2. The plotted data represent the actual temperature of the same cooling coffee in a room temperature of 72°F. Vary A, B, and C to find a function model for the data. Compare the model with your model from Question 4. Tell how the cooling-coffee data was adjusted for Plot2.

7. Suppose the same coffee cup was used on another day when room temperature was 70°F and the actual temperature of the coffee was only 180°F at time 0. Write an equation that you think would model the cooling-coffee temperature. Explain your choices for a, b, and c.

8. Switch from Plot2 to Plot3. Find a function model for the data. (*Hint:* Find values for a and c first.)

Asymptotes for Exponentials

Teacher Notes

Activity Objective

Students use the Transformation Graphing App to study the effects of changes in the values of a, b, and c, on the graph of the exponential function $y = ab^x + c$ and on its horizontal asymptote.

Time

- 40–50 minutes

Materials/Software

- Transformation Graphing App
- Programs: A2L82A, A2L82B
- Activity worksheet

Notes

- Review with students how to deselect and select a plot in either the Y= or STAT PLOT screens.
- To help verify answers in Question 3, students can trace along the graphs.
- In Question 8, after finding values for a and c, students may find it instructive to set Step = 0.01 to investigate values of b.

Answers

1. $y = 15(0.8)^x$. The base in the exponential function is between 0 and 1, so the function is decreasing as x gets large. $y = 0$; values of y are getting arbitrarily close to 0 as values of x get arbitrarily large.

2. Answers may vary. Sample: The asymptote is $y = c$, so as c changes, so will the asymptote.

3. $y = c; a + c$ **a.** $y = 4; 12$ **b.** $y = 2; 12$ **c.** $y = -5; 5$ **d.** $y = -4; 11$

4. Answers may vary. Sample: $y = 135(0.95)^x$. $y = 0$ (the amount room temperature is above room temperature); 135 (the temperature of the coffee above room temperature at time $x = 0$)

5. 207°F

6. Answers may vary. Sample: $y = 135(0.95)^x + 72$. Each coffee temperature value was increased by 72.

7. $y = 110(0.95)^x + 70$. Let $b = 0.95$ since the cooling substance is still coffee. Let $c = 70$, the room temperature, which is what y must approach as x gets large. Let $a = 110$, which is how much warmer than room temperature the coffee is at time $x = 0$.

8. Answers may vary. Sample: $y = 140(0.85)^x + 20$.

Name ________________ Class ________ Date ________

Asymptotes for Rationals

Activity 66

FILES NEEDED: Transformation Graphing App
Program: A2L92

The function $y = \frac{1}{x}$, sometimes known as the inverse function, is the parent for the family of rational functions, $y = \frac{a}{x - b} + c$. The parent function has the two axes as its asymptotes, as shown on page 485 of your textbook.

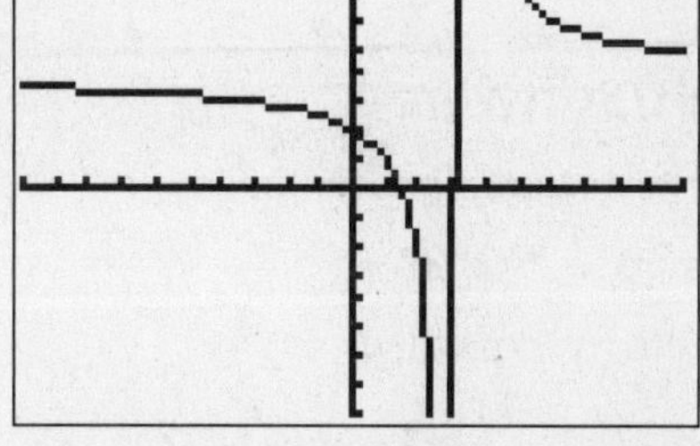

Because a rational function graph has two branches, it sometimes is difficult to get a good calculator view of the graph. On the other hand, if your calculator is in Connected Mode, the "connect" between the two branches (see right) will suggest the vertical asymptote!

In this activity, you will study the effects of changes in a, b, and c on the graphs of rational functions and particularly on their asymptotes. In fact, if you understand the effects of changes in the parameters on other families of functions, you already know how the changes will affect this family.

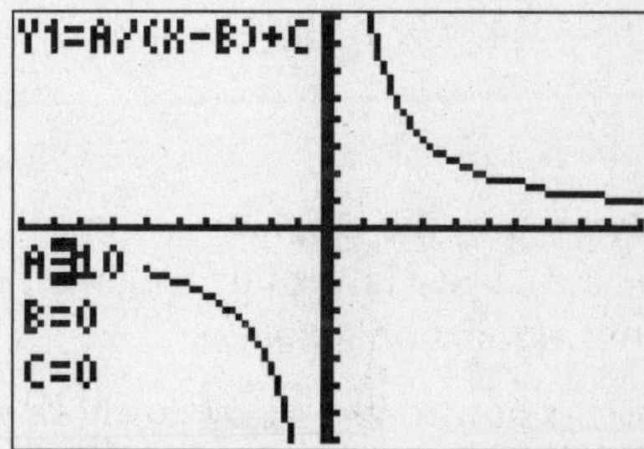

1. Run A2L92. Write the function equation for the graph shown on the startup screen. Compare its a value with that of the parent function. Tell how its graph compares to the graph of the parent function.

2. Predict how the graph will change when you change the value of B. Test your prediction with three different values for B. For each value, write the equations of the function and of both asymptotes.

3. Repeat Question 2 for C.

4. What are the equations of the asymptotes when B = -3 and C = 2?

5. Write equations for both asymptotes of each graph. Use A2L92 to check.

   a. $y = \frac{10}{x} + 2$ b. $y = \frac{10}{x - 2}$ c. $y = \frac{10}{x - 2} + 3$ d. $y = \frac{10}{x + 2} - 3$

6. Activate Plot1, Plot2, and Plot3, each in turn. For each graph, change parameter values in Y1 = A/(X − B) + C to make a matching graph. Write the function equation you find. Don't forget, you can change A.

   a. 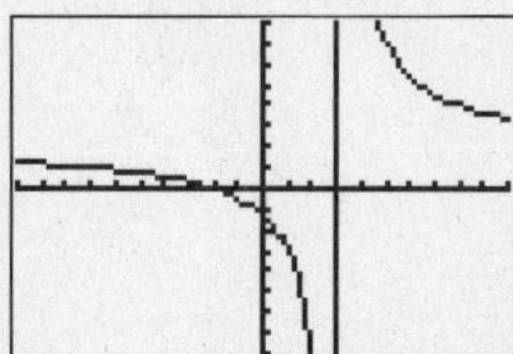 b. 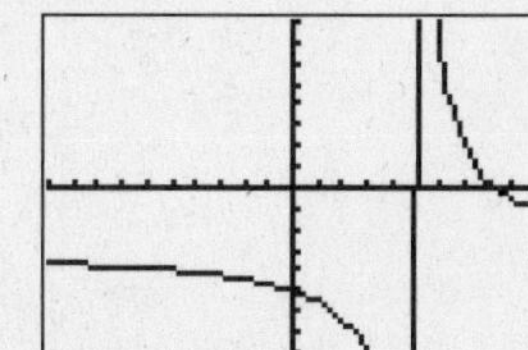 c. 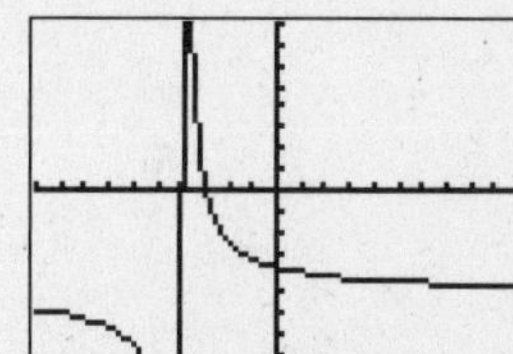

Asymptotes for Rationals

Teacher Notes

Activity Objective

Students use the Transformation Graphing App to study the connections between the values of b and c and the asymptotes of the rational function $y = \frac{a}{x - b} + c$.

Time

- 20–25 minutes

Materials/Software

- Transformation Graphing App
- Program: A2L92
- Activity worksheet

Skills Needed

- change parameter values
- select or deselect a plot

Notes

- Review how the graph of $y = a(x + b)^2 + c$ changes as a, b, and c change. Have students look for similar patterns for rational function graphs.
- Discuss how to use TRACE to check values for asymptotes.

Answers

1. $y = \frac{10}{x}$; for $y = \frac{10}{x}$, the value of a is 10. In the parent function, the value of a is 1. For both graphs the asymptotes are the x- and y-axes. The branches of $y = \frac{10}{x}$ are farther from the origin than the branches of $y = \frac{1}{x}$.

2. The graph will shift horizontally by an amount equal to the change in B.

3. The graph will shift vertically by an amount equal to the change in C.

4. $y = 2, x = -3$

5. **a.** $y = 2, x = 0$ **b.** $y = 0, x = 2$ **c.** $y = 3, x = 2$ **d.** $y = -3, x = -2$

6. **a.** $y = \frac{10}{x - 3} + 2$ **b.** $y = \frac{10}{x - 5} - 3$ **c.** $y = \frac{5}{x + 4} - 5$

Name ______________________ Class ______________ Date ______________

Sine Function Match

Activity 67

FILES NEEDED: Transformation Graphing App
Program: A2L134

A2L134 graphs the general form of the sine function $y = a \sin(b(x - c)) + d$, sometimes called a *sinusoid*, as Y1 = A sin(B(X − C)) + D.

In this activity you see three sine curves, one at a time. You are to change the values of A and B to find a perfect match for each given curve. Make sure that your graphing calculator is in Radian mode.

1. Run A2L134. What is the scale along the x-axis? Along the y-axis? Change values for A and B to find a sine function whose graph matches the given curve. Write the equation of this function.

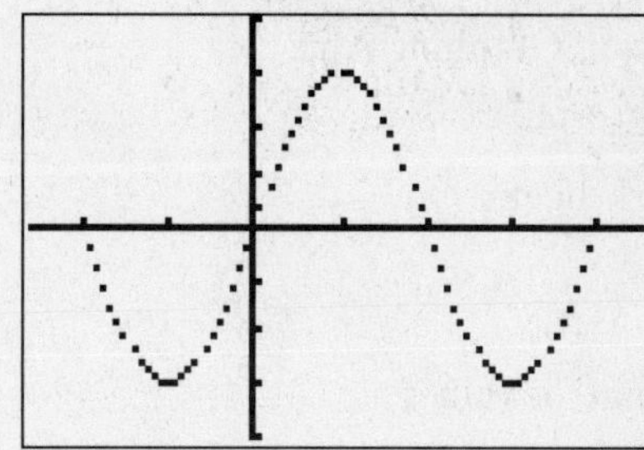

2. Switch from Plot1 to Plot2. Press GRAPH to see the second curve. Find a sine function whose graph matches the given curve. Then write the equation.

```
Plot1 Plot2 Plot3
Y1=Asin(B(X-C))
+D
Y2=
Y3=
Y4=
Y5=
Y6=
```

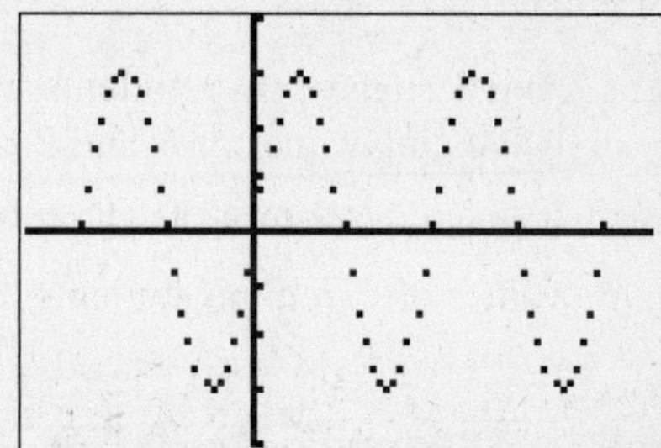

3. Switch from Plot2 to Plot3. Press GRAPH to see the third curve. Find a sine function whose graph matches the given curve. Write the equation.

4. Turn off all Plots. Set A = 2 and B = 1. Predict the effects that changes in C and D will have on the graph. (You've seen these effects before!) Test your predictions.

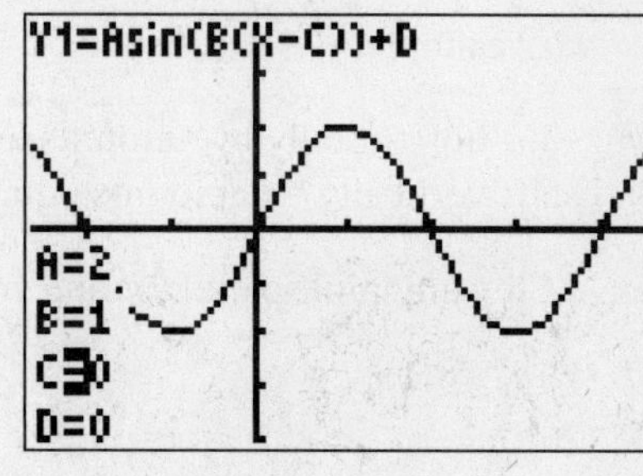

Extension

5. Use what you discovered in Question 4. Change the value of one of the parameters to translate the graph onto itself. List all values of the parameter for which you can do this.

Sine Function Match

Activity Objective

Students use the Transformation Graphing App to study the effects of changes in the values of a and b on the amplitude and wave length of the graph of the sinusoid $y = a \sin(b(x - c)) + d$.

Time

- 20–25 minutes

Materials/Software

- Transformation Graphing App
- Program: A2L134
- Activity worksheet

Skills Needed

- change parameter values
- select and deselect a plot

Notes

- You may need to remind students what it means for a scale to be in radians.
- For Questions 2 and 3, suggest that students try to match the functions by first changing values of A and B only.
- For Question 4, have students turn off all plots.
- When changing values of C, remind students to use multiples of π.

Answers

1. $\frac{\pi}{2}, \pi, \frac{3\pi}{2}, \ldots; 1, 2, 3, \ldots; y = 3 \sin x$
2. $y = 3 \sin 2x$
3. Answers may vary. Sample: $y = -1.5 \sin 0.5x$
4. The graph will shift horizontally by amounts equal to the changes in C. The graph will shift vertically by amounts equal to the changes in D.
5. You can change C by any multiple of 2π and translate the graph onto itself.

Name ____________________ Class ____________ Date ____________

Cosine Function Match

Activity 68

FILES NEEDED: Transformation Graphing App
Program: A2L137A

A2L137A graphs the general form of the cosine function $y = a \cos(b(x - c)) + d$ as Y1 = A cos(B(X − C)) + D.

In this activity you see three cosine curves, one at a time. You are to change the values of the parameters A, B, C, and D to find a perfect match for each given curve. Make sure that your graphing calculator is in Radian mode.

1. Run A2L137A. What is the scale along the x-axis? Along the y-axis? Change values of the parameters to find a cosine function whose graph matches the given curve. Write the equation of the function.

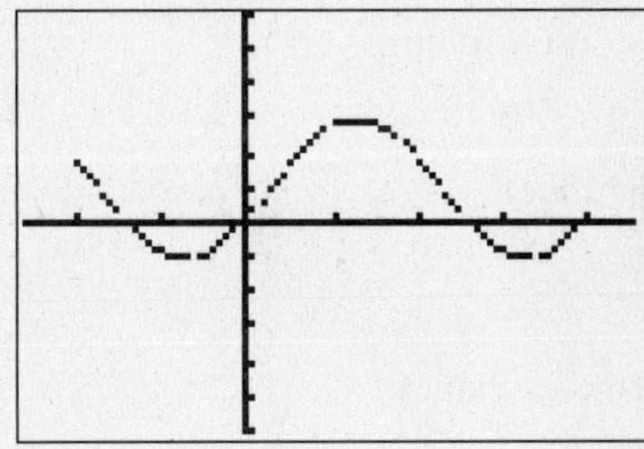

2. Switch from Plot1 to Plot2. Press GRAPH to see the second curve. Find a cosine function whose graph matches the given curve. Write the equation.

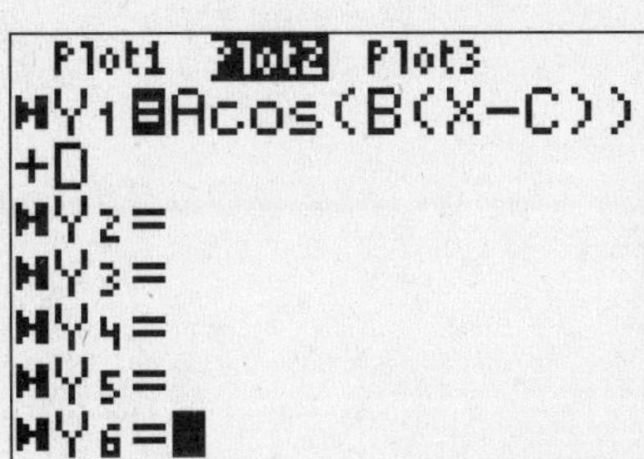

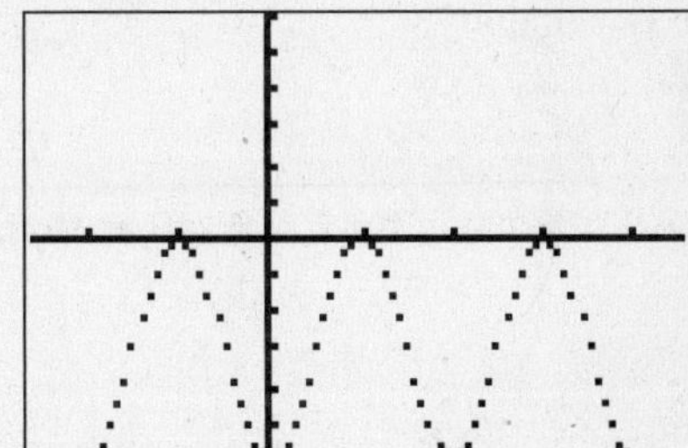

3. Switch from Plot2 to Plot3. Press GRAPH to see the third curve. Find a matching cosine function.

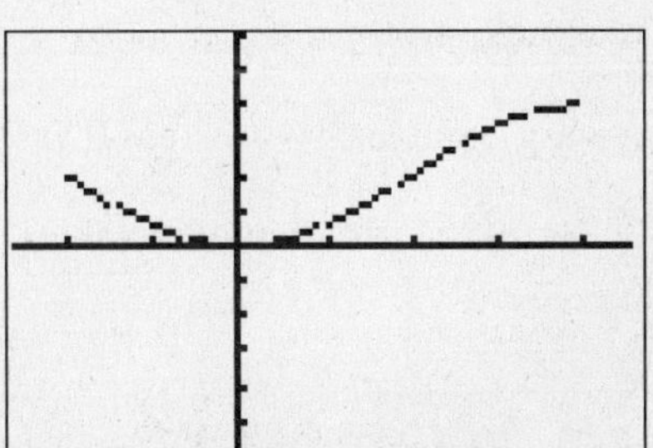

Extension

4. The sine function $y = \sin x$ and the cosine function $y = \cos x$ appear to be translations of each other. Show that this is true by finding the *phase shift* (value of c) that gives identical graphs for $y = \cos(x - c)$ and $y = \sin x$. List all such phase shifts.

Cosine Function Match

Activity Objective

Students use the Transformation Graphing App to study the effects of changes in the values of a, b, c, and d on the graph of the cosine function $y = a \cos (b(x - c)) + d$.

Time

- 25–30 minutes

Materials/Software

- Transformation Graphing App
- Program: A2L137A
- Activity worksheet

Skills Needed

- change parameter values
- select and deselect a plot

Classroom Management

- Students can work individually or in pairs depending on the number of calculators available.

Notes

- For Question 1, the value for C is 2 rather than a multiple of π.

Answers

1. $\frac{\pi}{2}, \pi, \frac{3\pi}{2}, \ldots; 1, 2, 3, \ldots; y = 2 \cos (x - 2) + 1$
2. Answer may vary. Samples: $y = -3 \cos 2x - 3$ or $y = 3 \cos (2x - \pi) - 3$
3. Answer may vary. Samples: $y = -2 \cos 0.5x + 2$ or $y = 2 \cos (0.5x - \pi) + 2$
4. $c = \frac{\pi}{2}; c = \frac{\pi}{2} + n(2\pi)$ for all integer values of n.

Name ______________________ Class ______________ Date ______________

Daylight Model

Activity 69

FILES NEEDED:	Transformation Graphing App Program: A2L137B

Run A2L137B. The startup screen plots hours of daylight in San Francisco, California, against the days of the year.

1. Write a short paragraph describing what happens to the number of daylight hours in San Francisco over one year.
2. Study the pattern of the plotted points. Of the functions you have studied so far—linear, absolute value, quadratic, radical, exponential, rational, and trigonometric—which do you think would make the best model for the relationship between the hours of daylight and the days of the year. Explain.
3. On the Y= screen, select the given sine function. (Highlight the "=" sign and press ENTER.) Press GRAPH. Change the values of the parameters A, B, C, D to find a sine function, $y = a \sin (b(x - c)) + d$, that is a good model for the data. Express A to one decimal place and B to three decimal places. When you feel you have a good model, zoom in to get a closer view. Refine your model if needed.
4. Use your model. In San Francisco, how long is the "longest day"? The shortest day? Today?
5. Activate both Plot1 and Plot2. Press GRAPH to see daylight data for both San Francisco and Anchorage, Alaska. Describe and explain the differences and similarities in the two plots.
6. Find a sine function model for the Anchorage data.
7. How much longer is the longest day in Anchorage than the longest day in San Francisco? How many days in Anchorage are longer than the longest day in San Francisco?
8. Predict differences you would expect to see in these plots and a plot for Sydney, Australia. Explain your predictions. Activate Plot3 and press GRAPH to see a plot for Sydney.
9. Why is the plot for Sydney "upside down"?
10. The data sets for the three cities "intersect" twice. Why? On what days does this happen? How long are these days?
11. In your sine models, what is the meaning of the value of D?

Extension

12. Use local resources and develop a model for your hometown. For your model, which of the parameters, A, B, C, and D would have the same values as for San Francisco, Anchorage, or Sydney?

Daylight Model

Teacher Notes

Activity Objective

Students use the Transformation Graphing App to make a sine-curve model of the hours of daylight over a year.

Time

- 40–50 minutes

Materials/Software

- Transformation Graphing App
- Program: A2L137B
- Activity worksheet

Skills Needed

- change parameter values
- select and deselect a plot

Notes

- Discuss with the students the use of the word *equinox.*
- Question 11 asks about the meaning of the value for D. You may wish to discuss why a good estimate for B is $\frac{2\pi}{365}$ and why a good estimate for C is 81 (the number of days from January 1 to March 22).

Answers

1. Answers may vary. Sample: The number of hours increase and then decrease.
2. Trigonometric; the plot suggests a sine function.
3. Answers may vary. Sample: $y = 2.6 \sin (0.017(x - 81)) + 12$
4. Answers based on the answer for Question 3. longest day: about 14.6 h; shortest day: about 9.4 h
5. Both sets of data follow a sine curve. Daylight hours for Anchorage are fewer in winter and greater in summer.
6. Answers may vary. Sample: $y = 6.8 \sin (0.017(x - 81)) + 12$
7. about 4.2 h; about 137 days
8. Sidney will have longer days when Anchorage and San Francisco have shorter ones.
9. Sidney is in the Southern Hemisphere
10. On about March 22 and September 22, all three cities have about 12 h of daylight.
11. D is the average length of a day, which is 12 h.
12. Check students' work; The parameter values for B, C, and D will be the same.